Student Companion

Tools and Techniques
for Individual and Group Learning for

C H E M I S T R Y

Molecules, Matter, and Change

FOURTH EDITION

LYNN GEIGER
BELIA STRAUSHEIN
LORETTA JONES
University of Northern Colorado

W. H. FREEMAN AND COMPANY
NEW YORK

ISBN 0-7167-3631-4

Copyright © 2000 by W. H. Freeman and Company

No part of this book may be reproduced by any mechanical, photographic, or electronic process, or in the form of a phonographic recording, nor may it be stored in a retrieval system, transmitted, or otherwise copied for public or private use, without written permission from the publisher.

Printed in the United States of America

First printing 1999

CONTENTS

Preface

1	Matter	1
2	Measurements and Moles	19
3	Chemical Reactions	37
4	Chemistry's Accounting: Reaction Stoichiometry	53
	Connection 1: Chemistry in the Drugstore	65
5	The Properties of Gases	69
6	Thermochemistry: The Fire Within	83
7	Atomic Structure and the Periodic Table	103
8	Chemical Bonds	121
9	Molecular Structure	139
	Connection 2: Finding Energy for the Future	153
10	Liquids and Solids	155
11	Carbon-Based Materials	171
12	The Properties of Solutions	189
	Connection 3: Sports Drinks	209
13	The Rates of Reactions	211
14	Chemical Equilibrium	227
15	Acids and Bases	239
16	Aqueous Equilibria	259
	Connection 4: What's in Our Water?	275
17	The Direction of Chemical Change	277
18	Electrochemistry	291
	Connection 5: Electric Vehicles	307
19	The Elements: The First Four Main Groups	309
20	The Elements: The Last Four Main Groups	319
21	The *d* Block: Metals in Transition	331
22	Nuclear Chemistry	343

PREFACE

To the instructor

An introduction to the use of cooperative learning in the classroom and a description of cooperative learning classroom tools is included in the *Instructor's Resource Manual.* Many of the tools discussed in the Manual are based on work done with Idahlynn Karre (*Busy, Noisy and Powerfully Effective: Cooperative Learning in the College Classroom*, Educational Videotape and Faculty Development Manual, Greeley, CO, University of Northern Colorado Research Corporation, 1996). For a copy of the videotape and/or faculty handbook, contact UNC Research Corporation, 800 8th Ave., Suite 140, Greeley, CO 80631.

To the student

This supplement is designed to help you succeed in general chemistry. It can be used to help you focus on specific concepts, to develop critical thinking skills, and to develop problem-solving strategies and skills in precise reading and interpretation of scientific data. The activities in this supplement were developed to provide additional assistance when working with your textbook and the CD-ROMs in conjunction with class activities. Not only do many activities have real-life applications, but an emphasis is placed on inquiry-based concept development. Our goal is to provide a way to help clarify chemistry concepts in a manner that has proven successful.

Each chapter has guided reading activities designed to help you focus on a specific topic or concept in the textbook chapter. The first component is an individual assignment that guides you through an important part of the chapter. This is followed by a group assignment in which you can develop methods to help solve additional problems. In addition, each chapter has an activity to help direct your thinking while viewing demonstrations or simulations on either the Problem-Solving Skills or Visualization CD-ROM. Some of these activities are incorporated into other assignments, while others are presented separately. Most chapters also have group activities that involve building models, solving challenge problems, and reviewing class material.

Our experience has proven that students who master these activities are also successful in their understanding of chemistry and therefore do well in their class. You will also learn social skills, such as group cooperation and leadership, that you will find useful in the workplace—no matter what career you choose.

CD-ROM Hints

There are several interesting components of the CD-ROMs that you will find useful in learning chemistry. Take the time to familiarize yourself with these features, as many of them are used in the *Student Companion* activities.

On the Problem-Solving Skills CD-ROM

Interactive periodic table
- found in the list both inside the main program (Chem 4 Skills) and outside the main program

Calculators
- can be reached outside the main program
- a scientific calculator containing an equilibrium problem-solver can be found outside of the main program

Equation plotting and curve-fitting programs
- found outside listed as individual tools (2D and 3D plotter, and curvefitter) of the main program

Demonstrations and simulations
- listed by chapter inside the main program (Chem 4 Skills)

Structures of molecules
- found outside of the main program under Chem 4 Molecules

Quizzes and flash cards
- to be used as review for in-class exams and quizzes
- found in the main program (Chem 4 Skills)

On the Visualization CD-ROM
- *Simulations and textbook animations*
- *Nomenclature review exercises*

Acknowledgments

No project as extensive as writing a book can be accomplished without the support of colleagues, friends, and families. Many people have encouraged us and we wish to thank some of them here. Loretta Jones and Lynn Geiger would like to extend a special thank you to Idahlynn Karre, University of Northern Colorado. Without her support and encouragement, none of this would have begun. She convinced us that cooperative learning was possible in lecture halls and spent many hours helping us to transform our classrooms into "busy, noisy, but powerfully effective" learning environments.

We would also like to thank some of the many people who provided us with ideas and feedback throughout our development work. At the University of Northern Colorado: Clark Fields, Lynn James, Nathaniel Cooper (now at Steamboat Springs High School), Julie Henderleiter (now at Grand Valley State University), Tom Pentecost, Peter Thomas (now at The University of Sciences and Arts of Oklahoma), and Kirk Voska (now at Truman State College); at the University of Illinois at Urbana–Champaign: Gilbert P. Haight, Jr. and Patricia Plaut; and at Strathclyde University: Alllison Littlejohn. Much of our development work was supported by RMTEC (Rocky Mountain Teacher Education Collaborative), a National Science Foundation-sponsored project that supports course revision in science and mathematics. We would especially like to thank our colleagues at Metropolitan State College of Denver, Colorado State University, Aims Community College, Front Range Community College, and the Community College of Denver. These individuals worked with us on chemistry curriculum reforms in the RMTEC project and provided greatly appreciated moral support. Finally, we would like to thank our families. Belia Straushein and Lynn Geiger would like to thank their husbands, Larry Straushein and Richard Schwenz, for all the times they watched the children while Belia and Lynn attended meetings and worked on the book. We would also like to thank our children for their patience and understanding: Ben and Katey Straushein, and Caroline and Robert Schwenz. Loretta Jones would like to thank her parents, Walter and Magdalen Lucek, and all her sisters and brothers for the opportunity to learn about the importance of cooperation at an early age.

Lynn Geiger
Belia Straushein
Loretta Jones

Chapter 1

Matter

Activity 1.1	The Scientific Method—Group Problem and Lecture Demonstration, p. 3	
Activity 1.2	Elements and Isotopes—Group Problem, p. 5	
Activity 1.3	Nomenclature—Guided Reading, p. 7	
Activity 1.4	Nomenclature—Worksheets, p. 11	
	Worksheet I	Binary Compounds and Binary Acids, p. 11
	Worksheet II	Stock Nomenclature, p. 12
	Worksheet III	Polyatomic Nomenclature—Salts and Acids, p. 13
	Worksheet IV	Extra Nomenclature Exercises—Salts and Hydrates, p. 14
	Worksheet V	Nomenclature Review, p. 15
Activity 1.5	Types of Matter—Group Problem, p. 17	

ACTIVITY 1.1
The Scientific Method
Group Problem and Lecture Demonstration

Name _____

Group _____

1. Form subgroups of 3 or 4 and record all of your thoughts and decision-making steps as you consider the solution of the following everyday problem:

    How would you go about finding a good part-time job to help pay for your college expenses?

    Solutions:

2. Rejoin your group and carefully watch the lecture demonstration. Record your group discussion below.

    a. Observations:

    b. Hypotheses:

    c. Additional experiments needed to test your hypotheses:

STUDENT COMPANION FOR CHEMISTRY

ACTIVITY 1.2
Elements and Isotopes
Group Problem

Name _____

Group _____

1. Assign each member in your group one of the following elements. Look up all of the naturally occurring isotopes for your element, record this information below, and bring this information to class. (A good reference is the *CRC Handbook*.)

   Elements: mercury, platinum, xenon, molybdenum, carbon, boron

   Element symbol and name:

Isotope symbol	Percentage abundance	Mass (atomic mass units)

2. In class: Pick several naturally occurring isotopes from the data that your group collected and use this information to fill out the following table.

Isotope symbol	Atomic number	Mass number	Number of p	Number of e^-	Number of n

STUDENT COMPANION FOR CHEMISTRY

ACTIVITY 1.3
Nomenclature
Guided Reading

Name _____

Group _____

Read Sections 1.8, 1.9, 1.10, and 1.11 in your textbook and answer the following questions.

1. Explain the difference between an element and a compound.

2. Give one example of an element and draw a picture of the smallest unit of your element, using a space-filling model.

3. Give one example of a compound and draw a picture of the smallest unit of your compound, using a space-filling model.

4. Describe the difference between an organic and an inorganic compound and give one example of each.

STUDENT COMPANION FOR CHEMISTRY

5. Define anion and cation. What types of elements form monatomic anions? What types of elements form monatomic cations?

6. Which group in the periodic table forms stable ions with the following charges?

   a. +1 _____   d. −3 _____

   b. +2 _____   e. −2 _____

   c. +3 _____   f. −1 _____

7. Explain the difference between a monatomic and a polyatomic ion. Give the name and formula for one example of each.

8. Work the exercises assigned by your instructor

Read Sections 1.15 through 1.17 in your textbook. Return to class and discuss the following questions in your group.

9. Why do chemists prefer to use the systematic name for a chemical instead of its common name? What important information is included in a systematic name that is not included in a common name?

10. Using the information in Sections 1.15 through 1.17, complete the following naming flow chart with the help of your group members. Use this flow chart to complete the naming worksheets or other assignments given by your instructor. Bring the chart to class and use it to help you with naming practice in class.

Naming Flow Chart

Read the questions on this sheet and then fill in the correct procedure or rule for naming a particular type of compound. Provide at least two examples for each type of compound in the spaces provided. Refer to your textbook for tables of common ions, polyatomic ions, prefixes, and how to name acids.

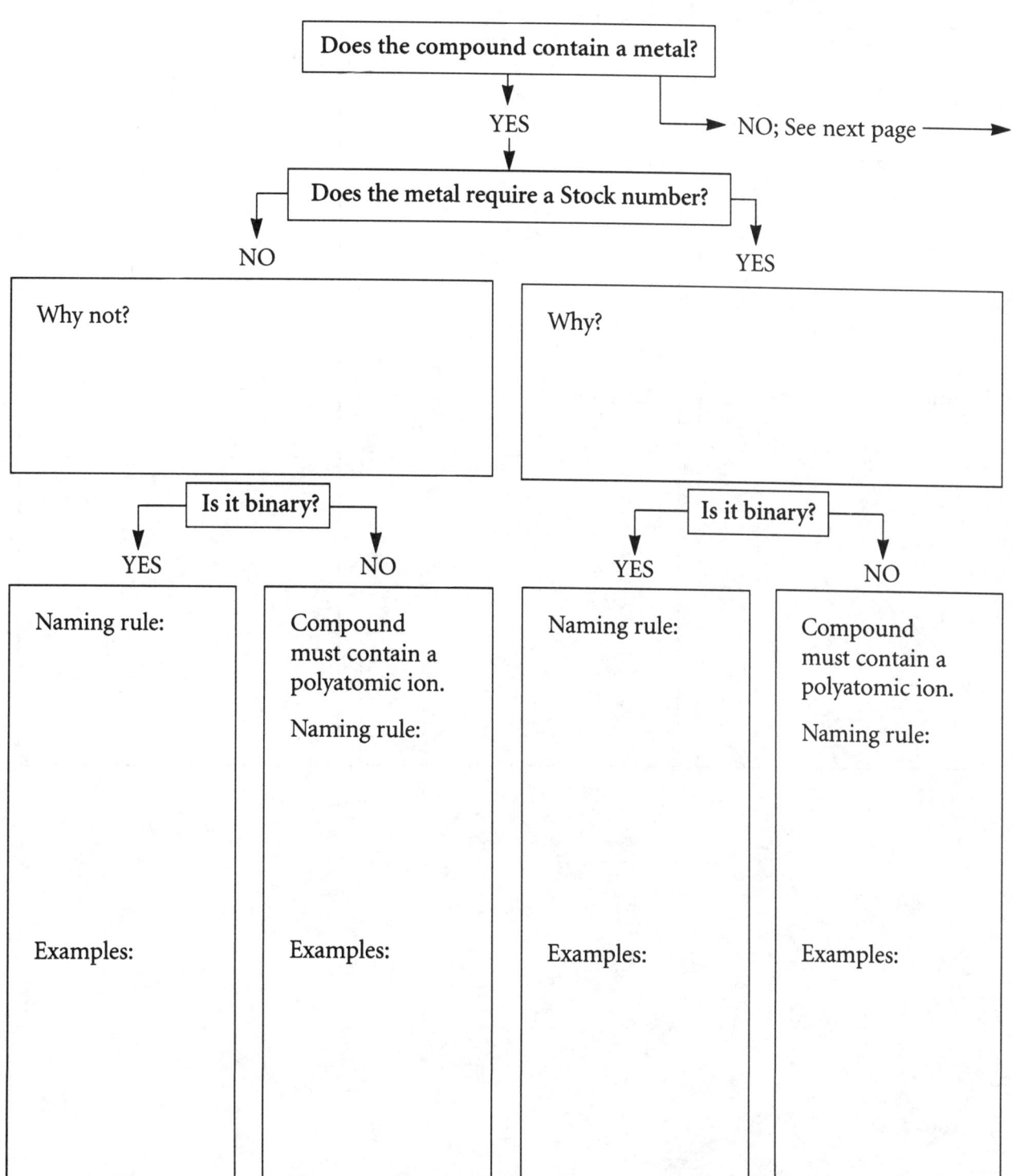

STUDENT COMPANION FOR CHEMISTRY

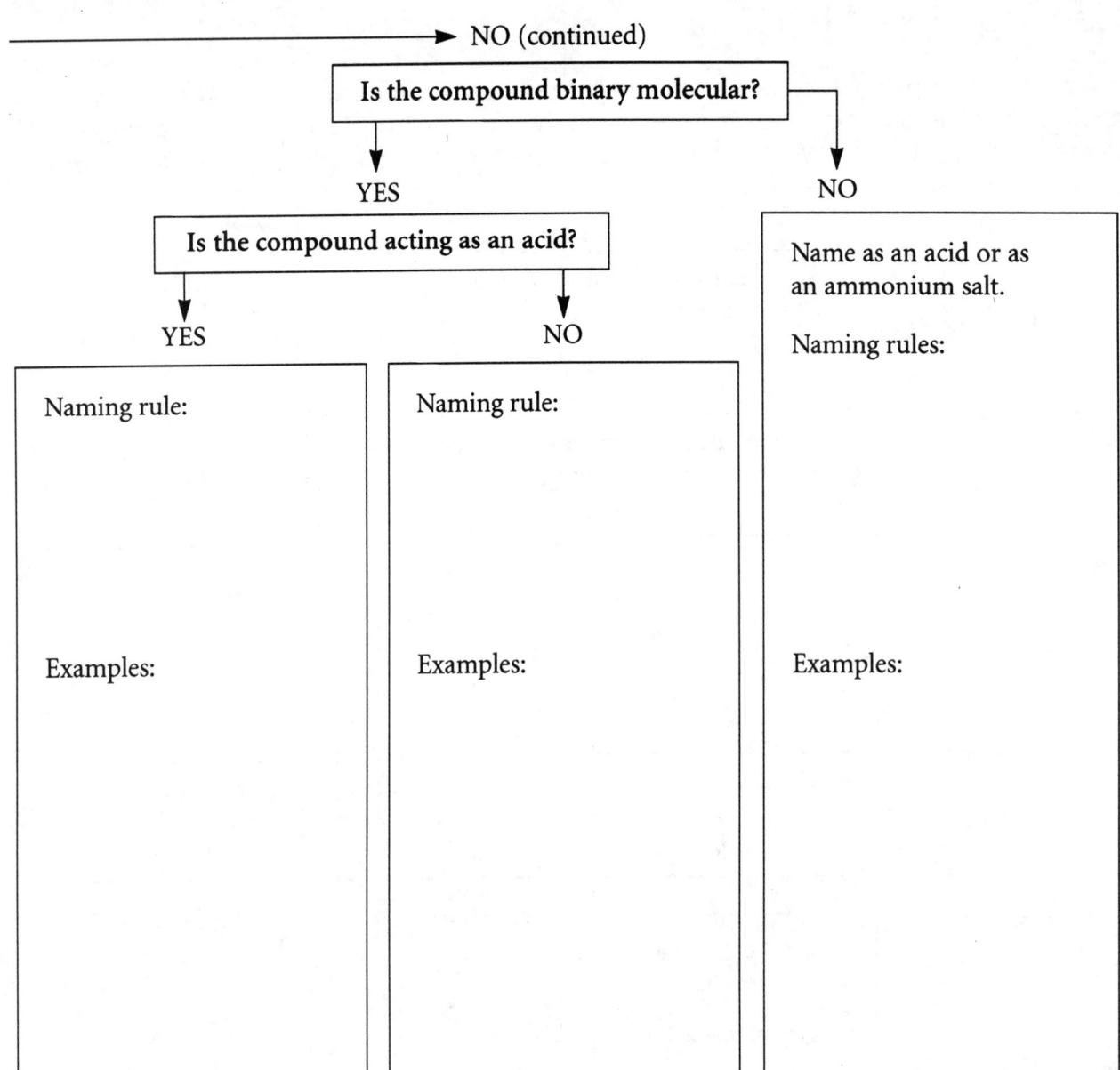

ACTIVITY 1.4
Nomenclature
Worksheets

Name _____

Group _____

Worksheet I   Binary Compounds and Binary Acids

1. Al_2O_3 _____
2. SiI_4 _____
3. AsF_3 _____
4. Cs_3N _____
5. SiO_2 _____
6. N_2O_5 _____
7. $LiCl$ _____
8. B_2S_3 _____
9. K_2O _____
10. PF_5 _____
11. BeO _____
12. $SeCl_2$ _____
13. I_2O_7 _____
14. BaI_2 _____
15. $HCl(aq)$ _____
16. MgS _____
17. $H_2S(aq)$ _____
18. Na_3N _____
19. OF_2 _____
20. $HI(aq)$ _____

sodium sulfide _____
calcium bromide _____
barium sulfide _____
phosphorus triiodide _____
sulfur trioxide _____
aluminum selenide _____
silicon disulfide _____
carbon dioxide _____
sodium oxide _____
carbon monoxide _____
lithium fluoride _____
boron trichloride _____
lithium phosphide _____
diarsenic pentoxide _____
aluminum oxide _____
hydrofluoric acid _____
strontium fluoride _____
potassium sulfide _____
cesium telluride _____
hydroselenic acid _____

STUDENT COMPANION FOR CHEMISTRY

Worksheet II   Stock Nomenclature

1. HgSO$_4$ _____
2. Cu$_3$P$_2$ _____
3. Sn$_3$(PO$_4$)$_2$ _____
4. Pb$_3$(PO$_4$)$_2$ _____
5. Cu(ClO$_3$)$_2$ _____
6. Fe$_2$(SO$_4$)$_3$ _____
7. Sn(SCN)$_2$ _____
8. Hg(NO$_3$)$_2$ _____
9. FePO$_4$ _____
10. PbO$_2$ _____
11. HgBr$_2$ _____
12. PbSO$_4$ _____
13. Hg$_2$F$_2$ _____
14. Fe$_2$(SO$_4$)$_3$ _____
15. HgO _____
16. CuI _____
17. Sn(SO$_4$)$_2$ _____
18. PbO _____
19. Cu(OH)$_2$ _____
20. Fe(CN)$_2$ _____

iron(II) sulfide _____
tin(II) oxalate _____
copper(II) oxide _____
iron(II) hydroxide _____
tin(IV) carbonate _____
mercury(II) oxide _____
copper(I) sulfite _____
copper(II) nitrate _____
lead(II) nitrite _____
lead(IV) iodide _____
iron(II) sulfide _____
mercury(I) oxide _____
copper(I) acetate _____
lead(II) phosphate _____
tin(II) phosphide _____
iron(III) permanganate _____
mercury(I) chloride _____
lead(II) fluoride _____
copper(II) nitrate _____
tin(IV) chloride _____

Name _____

Group _____

Worksheet III   Polyatomic Nomenclature—Salts and Acids

1. $Mg(ClO_3)_2$ _____
2. Na_2SO_4 _____
3. $Mg(CN)_2$ _____
4. Na_3PO_4 _____
5. $NaClO$ _____
6. Rb_2CO_3 _____
7. $HNO_2(aq)$ _____
8. $Cs_2C_2O_4$ _____
9. $BeCO_3$ _____
10. $HClO_3(aq)$ _____
11. $NaC_2H_3O_2$ _____
12. $BaSO_3$ _____
13. $Mg(NO_2)_2$ _____
14. $CaCO_3$ _____
15. $Al(OH)_3$ _____
16. $(NH_4)_2SO_4$ _____
17. $H_2SO_4(aq)$ _____
18. $Sr(ClO_3)_2$ _____
19. Na_2O _____

sodium cyanide _____
lithium chlorate _____
rubidium sulfate _____
ammonium thiocyanate _____
beryllium cyanide _____
calcium phosphate _____
oxalic acid _____
cesium carbonate _____
potassium dichromate _____
ammonium acetate _____
strontium chromate _____
sodium permanganate _____
magnesium phosphate _____
acetic acid _____
lithium peroxide _____
potassium nitrite _____
barium hydroxide _____
phosphoric acid _____
zinc hypochlorite _____

STUDENT COMPANION FOR CHEMISTRY

Name _____

Group _____

Worksheet IV   Extra Nomenclature Exercises

Salts with Hydrogen Anions

1. $NaHSO_4$ _____   calcium monohydrogen phosphate _____
2. $Ba(H_2PO_4)_2$ _____   ammonium bicarbonate _____
3. K_2HPO_4 _____   zinc dihydrogen phosphate _____
4. $LiHCO_3$ _____   cesium hydrogen sulfate _____
5. $Ca(HSO_3)_2$ _____   magnesium bicarbonate _____
6. NH_4HSO_4 _____   sodium hydrogen sulfite _____
7. $PbHPO_4$ _____   rubidium dihydrogen phosphate _____
8. $Sn(H_2PO_4)_2$ _____   barium monohydrogen phosphate _____

Hydrates

1. $SnCl_4 \cdot H_2O$ _____   zinc acetate dihydrate _____
2. $BaI_2 \cdot H_2O$ _____   ammonium phosphate trihydrate _____
3. $NaC_2H_3O_2 \cdot 3H_2O$ _____   sodium hypochlorite pentahydrate _____
4. $CuSO_4 \cdot 5H_2O$ _____   barium nitrite monohydrate _____
5. $NaIO_4 \cdot 3H_2O$ _____   calcium bromide hexahydrate _____
6. $HgSO_4 \cdot 2H_2O$ _____   iron(III) bromide hexahydrate _____
7. $Ca(ClO_2)_2 \cdot 3H_2O$ _____   lead(II) perchlorate trihydrate _____
8. $Na_3PO_4 \cdot 10H_2O$ _____   magnesium iodide octahydrate _____

Name _____

Group _____

Worksheet V   Nomenclature Review

1. K_2S _____
2. $HNO_3(aq)$ _____
3. Al_2O_3 _____
4. $Mg(NO_2)_2 \cdot 3H_2O$ _____
5. $Ca(IO)_2$ _____
6. P_4O_{10} _____
7. $H_2S(aq)$ _____
8. $H_2SO_3(aq)$ _____
9. $FeSO_3$ _____
10. $K_2C_2O_4$ _____
11. $Cu(IO_4)_2$ _____
12. $KMnO_4$ _____
13. $Ba_3(PO_4)_2$ _____
14. $HClO_4(aq)$ _____
15. SF_6 _____
16. $LiHCO_3$ _____
17. $HI(aq)$ _____
18. $Sn(BrO_3)_2$ _____
19. $NH_4C_2H_3O_2$ _____
20. $Sn(SCN)_4$ _____

lead(IV) perchlorate _____
sodium hypochlorite _____
perbromic acid _____
sodium oxalate _____
silicon tetraiodide _____
barium chloride _____
barium nitrite tetrahydrate _____
calcium chromate _____
copper(II) acetate monohydrate _____
hydrofluoric acid _____
nitrous acid _____
potassium iodide _____
lithium peroxide _____
hydrocyanic acid _____
phosphorus pentaflouride _____
acetic acid _____
aluminum chlorate _____
chloric acid _____
ammonium sulfide _____
aluminum dichromate _____

STUDENT COMPANION FOR CHEMISTRY

21. AgCl _____   magnesium hydroxide _____

22. Fe(HSO$_4$)$_2$ _____   lead(IV) bicarbonate _____

23. CuF$_2$ • 2H$_2$O _____   zinc sulfide _____

24. NaOH _____   strontium telluride _____

25. HClO(aq) _____   sodium sulfate _____

ACTIVITY 1.5
Types of Matter
Group Problem

Name Adelaida Tavarez

Group _____

9/4/01

1. What is a pure substance? Define these terms and explain the differences and similarities between mixtures and pure substances.

   A pure substance is something that is the same throughout, even on a microscopic scale. It is one form of matter, not a mixture. A mixture has different substances in it. It is a combination of many different thing to form itself. A mixture can be made up of different pure substances.

2. Explain the difference between a compound and an element.

   A compound is a substance that consists of 2 or more different elements with their atoms in a definate, characteristic ratio. An element is a substance composed of only one kind of atom.

3. Consider the gas state molecular pictures drawn below. Which picture is the best representation for a mixture? For an element? For a compound? Discuss your choices with the rest of your group.

   a.
   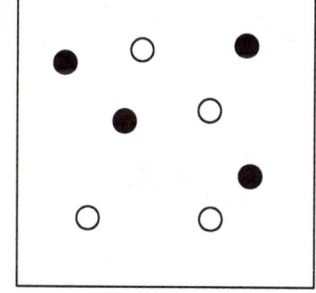

   b.
   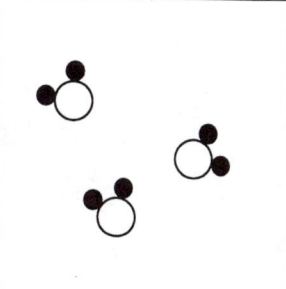

   c.
   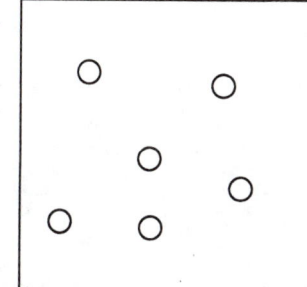

   Mixture — a
   Element — c
   Compound — b

STUDENT COMPANION FOR CHEMISTRY

17

4. What are the three states of matter.

solid
liquid
gas.

In the three boxes shown below, draw a molecular picture of water in each of its three states.

a. water vapor   b. liquid water   c. ice

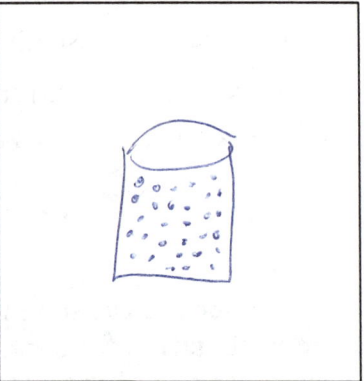

18   CHAPTER 1

Chapter 2

Measurements and Moles

Activity 2.1   Metric System and Density—Worksheet, p. 21
Activity 2.2   Significant Figures—Worksheet, p. 25
Activity 2.3   Empirical Formula—Guided Reading, p. 27
Activity 2.4   CD-ROM Demonstration—Viewing Guide, p. 31
Activity 2.5   Group Problem, p. 33
Activity 2.6   Group Challenge Problem, p. 35

ACTIVITY 2.1
Metric System and Density
Worksheet

Name __Adelaida Tavárez__

Group _____

1. Complete the following chart relating to SI prefixes.

Prefix	Symbol	Meaning	Decimal	Exponential
mega	M	million	1 000 000	1×10^6
kilo	K	thousand	1000	1×10^3
(unit)				
deci	d	tenth	0.1	1×10^{-1}
centi	c	hundreth	0.01	1×10^{-2}
milli	m	thousanth	0.001	1×10^{-3}
micro	µ	millionth	0.000001	1×10^{-6}
nano	n	billionth	0.000000001	1×10^{-9}
pico	p	trillionth	0.000000000001	1×10^{-12}

2. Compute the following conversions between units, using scientific notation.

a. 2789 m = __278,900__ / 2.789×10^5 cm

b. 63 998 kg = __63998000000__ / 6.3998×10^{10} mg

c. 42 ns = __.000000042__ / 4.2×10^{-8} s

d. 98.56 cg = __985.6__ / 9.856×10^2 mg

e. 76 455 Mm = __76455000000000__ / 7.6455×10^{12} mm

STUDENT COMPANION FOR CHEMISTRY

3. Write 3852 meters in terms of each of the following prefixes.

3.852×10^{-3} Mm $3852 \, m \times \dfrac{1 \, Mm}{10^6 \, m} = .003852$

3.852×10^{0} km $3852 \, m \times \dfrac{1 \, km}{10^3 \, m} = 3.852$

10^4 dm $3852 \, m$

10^5 cm

10^6 mm

10^9 μm

10^{12} nm

10^{15} pm

4. Write 654 meters squared as the square prefix of each of the following units.

10^{-4} km² $654 \, m^2 \times \dfrac{1 \, km^2}{(10^3)^2 \, m^2} =$

6.54×10^{2} m²

10^4 dm²

10^6 cm²

10^{15} mm²

Explain the strategy you used to determine your answers for Question 4.

22   CHAPTER 2

5. Write 3957 meters cubed as the cubic prefix of each of the following units.

   $\times 10^{-15}$ Mm³ $3.957 \, m^3 \times \dfrac{1 \, Mm^3}{(10^6)^3 \, m^3}$

   $\times 10^{3}$ m³

   $\times 10^{9}$ cm³

   $\times 10^{12}$ mm³

   $\times 10^{30}$ nm³

   Explain the strategy you used to determine your answers for Question 5.

6. Sometimes you will find that there are several ways to label a unit. Complete the following equivalences, which demonstrate this principle.

   For example, 0.123 g/mL = 0.123 g • mL⁻¹

   5.84 kJ/mol = 5.84 kJ • mol⁻¹

   1.3 g/cm³ = 1.3 g • cm⁻³

   10.2 mol/L = 10.2 mol • L⁻¹

7. Solve the following problems. Show your work to receive feedback on your responses.

   a. The largest nugget of gold on record was found in New South Wales, Australia, and had a mass of 94.5 kg. If the density of pure gold is 19.3 g/cm³, what was the volume of this pure nugget in cm³?

   b. What is the volume if the volume units were milliliters?

   c. If the market value of gold is $495 per ounce, how much is this nugget worth? (It may be helpful to use the resource inside the back cover of your text.)

STUDENT COMPANION FOR CHEMISTRY

8. A sample of ethanol with a volume of 46.0 mL and a density of 0.789 g/cm³ is added to a clean dry graduated cylinder that has a mass of 149.5 g. What is the mass of the cylinder after it has been filled with this volume of alcohol?

9. Mercury is a very dense material having a density of 13.545 g/cm³. What is the mass of 10.0 mL of pure mercury?

10. A piece of wood has a mass of 1648 g and dimensions of 173 cm by 6.8 cm by 5.5 cm. Calculate its density.

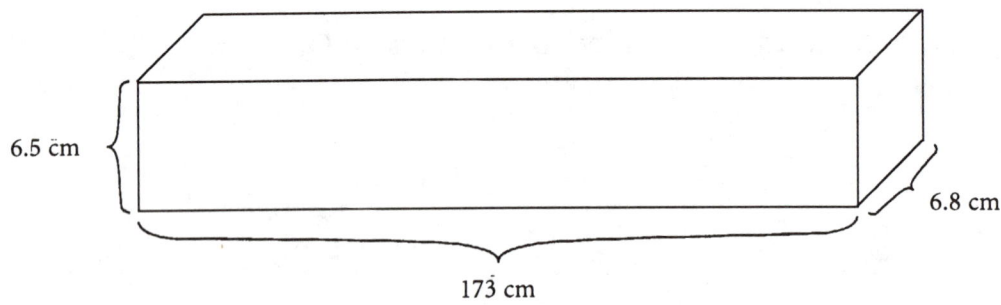

11. Discuss a situation in which the density of a substance might be a valuable piece of information.

24 CHAPTER 2

ACTIVITY 2.2
Significant Figures
Worksheet

Name __Adelaida Tavarez__

Group _____

1. Express the following numbers in scientific notation, showing the correct number of significant figures.

   a. 13 000 000 __1.3×10^7__

   b. 432 000 __4.32×10^5__

   c. 0.0004 __4.0×10^{-4}__

   d. 0.000 000 54 __5.4×10^{-7}__

   e. 73.744 __7.3744×10^1__

2. Solve the following problems, rounding off to show the correct number of significant figures.

   a. $73.98 + 3.5 + 0.01 =$ __77.49 = 77.5__

   b. $(1.4 \times 10^3) + (4.57 \times 10^2)(3.67 \times 10^1) =$ __1.8179×10^4__ ~~1.8×10^4~~

   c. $\dfrac{(0.591)(123.0)}{8.9} =$ __8.2__

   d. $\dfrac{(4.3 \times 10^2)(2.31 \times 10^4)}{(20.05)(9.1 \times 10^3)} =$ __5.4×10^1__ ~~54.4~~

3. Write the basic rule that applies to significant figures when two or more numbers undergo multiplication or division.

   You use the smallest # of significant figures in the data.

STUDENT COMPANION FOR CHEMISTRY

4. What is the basic rule that applies to significant figures when addition or subtraction occurs?

You use the smallest # of decimal places in the data.

5. A student working on a chemistry exam uses a calculator to determine an answer to a problem. The calculator display reads 56.990341. Discuss what circumstances are understood if this student writes the complete answer as shown.

If calculations require addition or subtraction that means the least # of decimal places was 6.

If it dealt with multiplication or division then at least 1 of the number would have to have at least 8 significant figures.

ACTIVITY 2.3
Empirical Formula
Guided Reading

Name _____ 9/25/01

Group _____

Read Sections 2.12 and 2.13 in your textbook. Complete the following questions individually.

1. Explain the difference between the terms *empirical* and *molecular* as they relate to chemical formulas.

   The empirical formula shows the relative # of atoms of each element and the molecular formula tells the actual #'s of atoms of each element in a molecule.

2. What information do you need to determine the empirical formula of a substance?

   The information you need is the mass percentage and the molar masses of the elements present.

3. When you know the empirical formula, what information is needed to determine the molecular formula?

   You need the compound's molar mass. You can calculate how many empirical formula units are needed to account for it.

STUDENT COMPANION FOR CHEMISTRY

4. Look at the following diagrams. Draw empirical formulas representing the molecular diagrams shown.

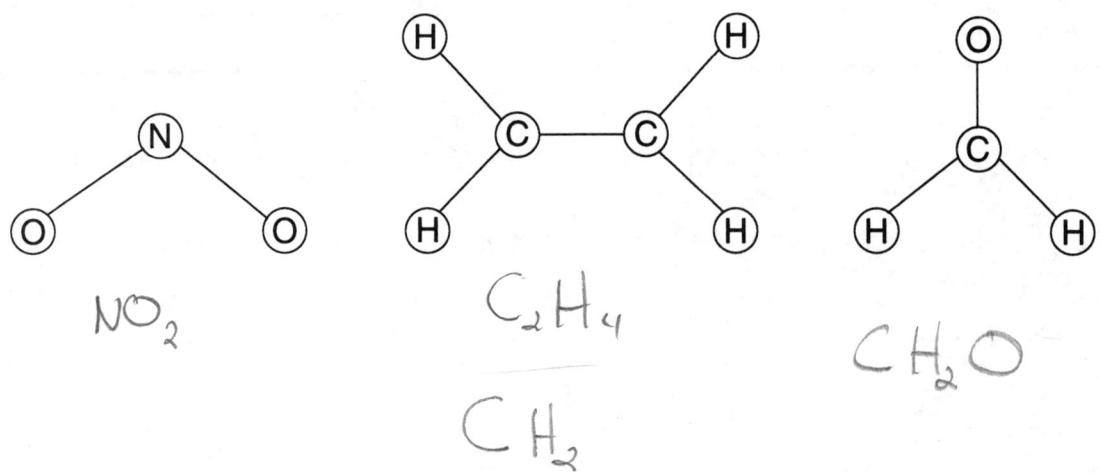

NO₂

C₂H₄

CH₂

CH₂O

5. Complete the homework problems assigned by your instructor.

Once you have completed the individual section of this assignment, work with your group to answer the following questions.

6. You are a laboratory technician who has been given a sample of a liquid substance found in a custodial closet and simply labeled "disinfectant." After analysis, your group determined that the active ingredients contained 33% sodium, 36% arsenic, and 31% oxygen by mass.

   a. Develop a group strategy for finding the empirical formula of this disinfectant.
   - 100g sample
   - convert % to g divide by atomic mass

   b. Use your strategy to determine the empirical formula of the disinfectant.

   sodium - 22.99
   arsenic - 74.92
   oxygen - 16

   33g Na × $\frac{1 \text{ mol}}{22.99 \text{ g}}$ = 1.44 mol Na

   36g As × $\frac{1 \text{ mol}}{74.92 \text{ g}}$ = .48 mol As

   31g O × $\frac{1 \text{ mol}}{16 \text{ g}}$ = 1.93 mol O

   $\frac{1.44}{.48}$ = Na
   $\frac{.48}{.48}$ = As
   $\frac{1.93}{.48}$ = O

   1.93 28

   Na₃AsO₄

CHAPTER 2

7. Without calculating, predict which of the following compounds has a larger percentage of magnesium, MgO or MgSO$_4$? Explain your reasoning.

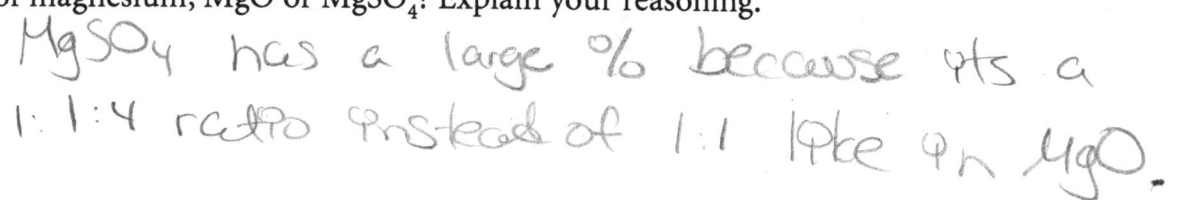

MgSO$_4$ has a large % because its a 1:1:4 ratio instead of 1:1 like in MgO.

8. Your group designs an experiment to determine the empirical formula of nicotine in cigarette smoke. Your results show that in a 100.0-mg sample of smoke, there are 74.0 mg of carbon, 8.7 mg of hydrogen, and 17.2 mg of nitrogen.

   a. On the basis of this information, what empirical formula do you calculate for nicotine?

   $\frac{mg}{1000} = .074 g$

   $\frac{7 mg}{1000} = .0087 g$

   $\frac{17.2 mg}{1000} = .0172 g$

   $.074 g\, C \times \frac{mol}{12.01 g} =$

   $.0087 g\, H \times \frac{mol}{2.0} =$

   $.0172 g\, N \times \frac{mol}{\_\_} =$

   b. What information would you need to determine the molecular formula of nicotine?

   You need the molar mass.

9. Isooctane is one of the most common components of gasoline. It is a hydrocarbon comprising hydrogen and carbon, with a molecular mass of 114.2 g/mol. A sample is found to have 15.9% hydrogen and 84.1% carbon by mass. What is this hydrocarbon's molecular formula?

100 g sample
H = 15.9 g → *convert to mol = 15.9 mol*
C = 84.1 g → *"   " =*

STUDENT COMPANION FOR CHEMISTRY 29

ACTIVITY 2.4
CD-ROM Demonstration
Viewing Guide

Name _____

Group _____

View the demonstration "Reaction of Sodium and Chlorine" on the Problem-Solving Skills CD-ROM.

1. What are the two components of a binary ionic compound?

2. Write a balanced chemical equation for the formation of sodium chloride as described in the demonstration.

3. Discuss how a neutral atom forms a cation.

4. Discuss how an anion forms from a neutral atom.

STUDENT COMPANION FOR CHEMISTRY

5. Explain how cations and anions are held together in a compound, using sodium chloride as your example. Include in your discussion a sketch of the molecular arrangement of ions.

6. If you produced 175.5 g of NaCl, how many moles of sodium chloride would you have?

7. Using your periodic table and Question 6 as a guide, write your own problem in which the product is a binary ionic compound other than sodium chloride. Provide information to calculate a product yield. Then, solve this problem.

ACTIVITY 2.5
Group Problem

Name _____

Group _____

Solve the following problems with your group. Your signature at the top of this sheet indicates your participation and agreement with group answers.

Pure ethanol (C_2H_5OH) is the ingredient measured and reported as its "proof" on the label of a bottle of alcohol in the United States. When alcohol was produced during Prohibition, consumers demanded that the maker demonstrate that the substance was actually alcohol. This was done by exposing the alcohol to fire. If the alcohol vapors ignited, it was considered to be "proof." The term *proof*, by definition, is twice the percentage by volume of pure ethanol in solution. This measurement is usually taken at 60°F. The density of ethanol at 60°F is 0.80 g/cm³ and the density of water at this same temperature is 1.0 g/cm³.

a. What is this recorded temperature in degrees Celsius?

b. Does density play a role in this problem? Explain.

c. If a sample of ethanol were determined to be "80 Proof," what is the percentage of water by volume?

d. Explain how to make 500 mL of this solution at this concentration. Assume that the final volume is equal to the original volumes of water and alcohol. Be precise in your quantities, but you may use the general terms of solute and solvent if you wish.

e. How many moles of ethanol are present in the solution?

STUDENT COMPANION FOR CHEMISTRY

ACTIVITY 2.6
Group Challenge Problem

Name _____

Group _____

1. Because you have demonstrated competence in your field, a contractor hires your group to find important information about a useful chemical compound. Initial testing shows that the compound contains 48% oxygen, 34% sodium, and 18% carbon by mass. The molar mass has been determined to be 134 g/mol.

   a. What are the empirical and molecular formulas for this compound?

   b. What is the systemic name of this compound likely to be?

   c. If 33.5 of this substance were dissolved in enough water to make 500.0 mL, what would the molarity of the solution be?

   d. Because this concentration is not what the lab wants, your group decides to dilute the solution by removing 1.0 mL of this solution (from c) and adding it to 9.0 mL of water to make a total of 10.0 mL of solution. What is the molarity of this new solution?

   e. Calculate the mass of solute (in grams) of this 10.0 mL of solution.

f. In the space below, sketch out a label to place on the tube of diluted solution that would have all of the necessary information needed on a chemical label.

Chapter 3

Chemical Reactions: Modifying Matter

Activity 3.1	Balancing Chemical Equations, Using Molecular Models—Group Hands-On Exercise, p. 39
Activity 3.2	Electrolytes—Group Problem and Demonstration, p. 41
Activity 3.3	Predicting Reaction Products—Group Problem and Demonstration, p. 43
Activity 3.4	Predicting Reaction Products and Writing Chemical Equations—Worksheet, p. 45
Activity 3.5	Oxidation-Reduction—Guided Reading, p. 47
Activity 3.6	Oxidation Numbers—Worksheet, p. 51

ACTIVITY 3.1
Balancing Chemical Equations, Using Molecular Models
Group Hands-On Exercise

Name _____

Group _____

Read Sections 3.1 and 3.2 in your textbook before coming to class.

Materials: Model kit

1. Consider the following questions.

   a. What are the mechanics of balancing a chemical equation? Describe how you balance an equation.

   b. Why do we balance chemical equations?

2. Methane is an organic molecule that is easily burned. Build a model and write a chemical formula for methane (methane has four hydrogens bonded to a central carbon atom). Save this model to use in Question 4.

STUDENT COMPANION FOR CHEMISTRY

3. a. When a substance is burned in air, the chemical reaction is called a combustion reaction. In the combustion reaction of methane in air what is the other reactant?

   b. Build four molecules of this element, and save them for Question 4.

4. The products of the combustion reaction of methane are carbon dioxide and water. Use the models that you made in steps 2 and 3b to build these products.

   a. How many carbon dioxide molecules can you build?

   b. How many water molecules can you build?

   c. Did you have any leftover reactant molecules? Which ones and how many?

5. Using the information that you gathered from 1, 2, and 3, write the balanced chemical equation for the combustion of methane. Write this equation two ways, using chemical formulas and drawing pictures of the reactant and product molecules. Do you need to include the leftover reactant molecules in your equations?

ACTIVITY 3.2
Electrolytes
Group Problem and Demonstration

Name _____

Group _____

1. Discuss with your group members the definitions of the following terms, and record your group's answers.

   a. strong electrolyte

   b. weak electrolyte

   c. nonelectrolyte

2. Watch the demonstration and fill in the following table. Decide whether each substance acts as a strong electrolyte, a weak electrolyte, or a nonelectrolyte and check the appropriate box. All solutions are aqueous.

Substance	Solute	Strong electrolyte	Weak electrolyte	Nonelectrolyte
a. salt (NaCl) water				
b. household ammonia				
c. hydrochloric acid				
d. rubbing alcohol				
e. baking soda in water				
f. sodium hydroxide				
g. sugar water				
h. tap water				

3. Draw pictures illustrating the molecules/ions in solutions a, b, and c as they exist in aqueous solution.

ACTIVITY 3.3
Predicting Reaction Products
Group Problem and Lecture Demonstration

Name _____

Group _____

1. Your instructor will mix two different aqueous solutions. Watch the lecture demonstration and record your observations below.

Observations	Reactants (formula and name)	Products (formula and name)
a.		
b.		
c.		
d.		

2. Write balanced molecular and net ionic equations for all reactions observed. Be sure to label states/phases of all reactants and products. If no reaction occurred, write NR.

   a. Molecular:

   Net ionic:

   b. Molecular:

   Net ionic:

   c. Molecular:

   Net ionic:

   d. Molecular:

   Net ionic:

STUDENT COMPANION FOR CHEMISTRY

ACTIVITY 3.4
Predicting Products
Worksheet

Name _____

Group _____

Complete and balance the following reactions. Write complete and net ionic equations for all reactions in solution.

Easier Equations

1. KOH(aq) + HNO$_3$(aq) →

2. Na$_2$CO$_3$(aq) + BaCl$_2$(aq) → 2NaCl(aq) + BaCO$_3$

3. Na$_2$SO$_4$(aq) + BaCl$_2$(aq) → 2NaCl + BaSO$_4$

4. Na$_3$PO$_4$(aq) + CuBr$_2$(aq) → Na$_3$Br$_2$ + Cu$_3$(PO$_4$)$_2$

5. Ba(OH)$_2$(aq) + HCl(aq) →

6. NaCl(aq) + K$_2$SO$_4$(aq) → Na$_2$SO$_4$(aq) + 2KCl(aq)

7. Ca(OH)$_2$(aq) + HC$_2$H$_3$O$_2$(aq) →

8. Sr(C$_2$H$_3$O$_2$)$_2$(aq) + NiSO$_4$(aq) →

9. AgNO$_3$(aq) + Na$_2$CO$_3$(aq) →

10. Ba(OH)$_2$(aq) + Fe$_2$(SO$_4$)$_3$(aq) →

11. Cr(OH)$_3$(s) + HNO$_3$(aq) →

12. Pb(NO$_3$)$_2$(aq) + K$_2$S(aq) →

13. Fe(OH)$_3$(s) + HClO$_4$(aq) →

More Challenging Equations

14. $CaCO_3(aq) +$ $HNO_3(aq) \rightarrow$

15. $CH_3OH(l) +$ $O_2(g) \rightarrow$

16. $HI(aq) +$ $NH_3(aq) \rightarrow$

17. $HNO_3(aq) +$ $NH_3(aq) \rightarrow$

18. $Na_2CO_3(aq) +$ $HBr(aq) \rightarrow$

19. $C_4H_{10}(g) +$ $O_2(g) \rightarrow$

20. $H_2SO_4(aq) +$ $Mg(HCO_3)_2(aq) \rightarrow$

ACTIVITY 3.5
Oxidation-Reduction
Guided Reading   10/11/01

Name _____

Group _____

Read the textbook's introductory section on redox reactions: Section 3.14. Answer the following questions.

1. What is the definition of a redox reaction?

   A reaction in which oxidation and reduction occur.

2. Define oxidation.

   Oxidation is electron loss, even if no oxygen is involved.

3. Define reduction.

   Reduction is electron gain.

4. Is it possible to have a reaction in which only oxidation or only reduction occurs? Explain your answer.

   No because every oxidation reaction must also involve a reduction reaction, and vise-versa.

STUDENT COMPANION FOR CHEMISTRY

Read Sections 3.11 and 3.12 in your textbook.

5. Define oxidation number.

the oxidation # keeps track of electrons of each element. An increase in the # indicates oxygen and a decrease in the # indicates reduction.

6. Assign oxidation numbers for all the elements in the following species.

a. H_2 = 0 $Na_{(s)}$ = 0

b. ClO_4^- = −1 OF_2 = −2 for O (+1; −2, +2)
(+7, −8)

c. C_2H_2 = 0 $C_2O_4^{2-}$ = +3 for C (+3, −2; +6, −8)
(−2, +2)

d. $MgSO_4$ = 0 XeO_4^{4-} = +8 for Xe (+8, −12)
(+2, +6, −8)

                 $MgCl_2$ = +2 for Mg (+2, −2)

7. Explain the difference between an oxidizing agent and a reducing agent. Which substance is oxidized and which is reduced? How would you identify these substances in a chemical equation?

Read Section 3.17 in your textbook and work the homework problems assigned by your instructor. Return to class and consider the following questions for class and group discussion.

8. Give some examples/uses of redox reactions.

48   CHAPTER 3

9. Consider the thermite reaction (lecture demonstration or Problem-Solving Skills CD-ROM demonstration), which is an exothermic reaction of iron(III) oxide with powdered aluminum.

   a. Look up the definition of exothermic and record it here.

   b. What do you think the products of this reaction are? Write a balanced chemical equation.

   c. Which substance in this reaction is oxidized? reduced? the oxidizing agent? the reducing agent?

STUDENT COMPANION FOR CHEMISTRY 49

ACTIVITY 3.6
Oxidation Numbers
Worksheet

Name _____

Group _____

10/16/01

Assign oxidation numbers for the following atoms.

1. B in BF_3 (F: -1) = +3
   +3 -3

2. C in CH_4 (H: +1) = -4
   -4 +4

3. N in NH_3 (H: +1) = -3
   -3 +3

4. P in PCl_3 (Cl: -1) = +3
   +3 -3

5. C in CO_2 (O: -2) = +4
   +4 -4

6. S in SO_2 (O: -2) = +4
   +4 -4

7. C in CH_2O (H: +2, O: -2) = 0

8. S in SF_4 (F: -1) = +4
   +4 -4

9. S in SF_6 (F: -1) = +6
   +6 -6

10. I in IF_5 (F: -1) = +5
    +5 -5

11. P in $Zn_3(PO_4)_2$

12. Cr in $K_2Cr_2O_7$ (K: +1, O: -2) = +6
    +2 +12 -14

13. Br in $HBrO_3$ (H: +1, O: -2) = +5
    +1 +5 -6

14. V in VO_2^- (O: -2) = +3
    +3 -4

15. Mn in MnO_4^- (O: -2) = +7
    +7 -8

16. N in NO_2 (O: -2) = +4
    +4 -4

17. N in N_2O_4 (O: -2) = +4
    +8 -8

18. N in N_2O_3 (O: -2) = +3
    +3 -6

19. As in H_3AsO_4 (H: +1, O: -2) = +5
    +3 +5 -8

20. P in $H_2PO_4^-$ (H: +1, O: -2) = +5
    +2 +5 -8

STUDENT COMPANION FOR CHEMISTRY

Chapter 4

Chemistry's Accounting: Reaction Stoichiometry

Activity 4.1   Stoichiometry—Demonstration, p. 55
Activity 4.2   Limiting Reactant—Guided Reading, p. 57
Activity 4.3   Group Problem, p. 59
Activity 4.4   Group Challenge Problem, p. 61
Activity 4.5   Titration—Worksheet, p. 63

ACTIVITY 4.1
Stoichiometry
Demonstration

Name _____

Group _____

1. Develop your own step-by-step strategy for solving problems in which you need to determine the limiting reactant in a reaction. (*Consider:* How do you determine how much product you can make or how much of other reactants are left over?)

2. Write a balanced chemical equation for the demonstration you see.

3. Complete the following chart. Use the cards for help.

Question	Flask 1	Flask 2	Flask 3	Flask 4
Mass of magnesium				
Volume of the solution				
Concentration of the solution				
Observations				
Which reactant is limiting?				
Which reactant is in excess? How much is left over?				
How many moles of H_2 are produced?				
How many grams of magnesium chloride are produced?				
Molar conc. of magnesium chloride at the end?				

ACTIVITY 4.2
Limiting Reactant
Guided Reading

Name _____

Group _____

Read Section 4.4 in your textbook. Complete the following questions without group discussion.

1. Discuss what is meant by the term *limiting reactant* in a chemical equation.

2. How do you use a chemical equation to determine the quantities of each reactant needed to produce a required amount of product?

3. Assume you are given an equation that contains two reactants. Describe the steps you can take to determine which reactant is the limiting reactant.

STUDENT COMPANION FOR CHEMISTRY

4. Rust is a complex mixture of iron oxides, one of which is iron(III) oxide. Write a balanced equation that shows the formation of rust. When a piece of metal rusts, what reactant is likely to be the limiting reactant? the reactant in excess?

5. Which reactant, the limiting reactant or the reactant in excess, is used to calculate the theoretical yield of a product? Explain how this is determined.

6. Complete the homework problems assigned by your instructor.

Complete the remainder of these problems with your group members.

7. Chlorine gas (Cl_2) is produced by passing an electric current through a concentrated sodium chloride solution known as brine. The sodium chloride in water solution, called aqueous brine, is then exposed to the current. The resulting products include sodium hydroxide and hydrogen gas in addition to chlorine gas. Suppose you had 1.5 kg of 24% brine (by mass), how much of each product would you be able to form?

8. Assume that you add an aqueous solution of iron(III) sulfate to a solution with barium chloride. The compounds react to produce barium sulfate, which precipitates to the bottom of the flask, and iron(III) chloride, which remains in solution.

   a. If you have 10.0 g of iron(III) sulfate, how much barium chloride would be needed to react with the iron(III) sulfate?

   b. If you have 50.0 g of iron(III) sulfate and twice that amount of barium chloride, how much iron(III) chloride could be produced? Is there a reactant in excess in this problem? If so, what is it and how much would remain?

ACTIVITY 4.3
Group Problem

Name _____

Group _____

Solve the following problem with your group. Your signature at the top of this paper indicates your participation and agreement with group answers.

1. A chemical compound used in photography is silver bromide. A photographer decides to mix her own developing solution by combining silver nitrate and sodium bromide. She needs to produce 20 lb of this silver bromide to replenish her supply.

   a. Make a list of the information you need to determine how mush of each reactant is needed to react completely to produce 20 lb of product.

   b. Write a balanced equation for this reaction.

   c. How many moles of *each* reactant is required to produce one mole of product?

STUDENT COMPANION FOR CHEMISTRY

d. One pound is equal to 454 g. How many grams of product does she need to produce?

e. What mass of each reactant, in grams, is required for this reaction?

ACTIVITY 4.4
Group Challenge Problem

Name _____

Group _____

Your signature at the top of this page indicates that you were an active participant in your group and that you agree with group answers.

1. A new fuel composed primarily of butene, C_4H_8, is being tested as a possible replacement fuel for special combustion engines. A full fuel container can hold 8.40 kg of this fuel. Preliminary testing is occurring to determine the efficiency of this fuel.

   a. Write a balanced equation that demonstrates the complete combustion of C_4H_8 to carbon dioxide gas and water.

   b. How many kilograms of each product could theoretically be produced from a full tank of this fuel?

   c. Early testing shows that a full tank of fuel produces 22.0 kg of carbon dioxide gas. Is this 100% of the theoretical yield?

   d. Combustion reactions typically do not produce 100% of the theoretical yield in most engines. As a result, other products also may be produced. What are these products? (*Hint:* Consider the elements that are found in the reactants; what different combinations could you make to predict possible products?)

STUDENT COMPANION FOR CHEMISTRY

e. If you were in charge of determining whether this fuel would be worth investigating further, what would your recommendation be? Use the space below to discuss your reasoning.

ACTIVITY 4.5

Titration
Worksheet

Name _____

Group _____

1. Diagram a titration setup. Label the following items on your sketch:

    a. titrant   b. analyte   c. buret   d. indicator

2. Refer to Toolbox 4.6, "How to interpret a titration." Answer the following questions.

    Step 1   What information is important to start with? What do you need to know in order to begin calculating?

    Step 2   Where do you find out the stoichiometric relationship that exists? What constitutes the conversion factor?

    Step 3   How do you know when the stoichiometric point has been reached in a titration? Where does this information fit in the calculation?

STUDENT COMPANION FOR CHEMISTRY

3. A student wants to know the concentration of an unknown sample of acetic acid (CH$_3$COOH) found in the back of his refrigerator in a container labeled salad dressing. He remembers from his chemistry lab that if you titrate this unknown acid with a known base, you can determine the concentration of the acid. He decides to use a concentration of 0.100 M NaOH as his base. Careful titration of a 50.0-mL sample of acid shows that it takes 26.5 mL of NaOH solution to reach an end point. Of course, he repeats his titration to make sure his data are correct and achieves the same end point value.

a. Write a balanced equation for this reaction. (Why is this important?)

b. Complete Step 1 from the Toolbox.

c. Complete Step 2 from the Toolbox.

d. Complete Step 3 from the Toolbox.

4. What was the concentration of the acetic acid in the container?

CONNECTION 1
Chemistry in the Drugstore

Name _____

Group _____

Early medicines were typically based on available plant products, and some of the earliest recorded uses were found in papyrus from 1550 BC. Since then, a lot of research and energy has been devoted to isolating specific chemical compounds and determining a way to manufacture them in large quantities. Some of the most widely used examples are pain relievers, antibiotics, and vitamins. Every year many new drug compounds are made available to the public. Testing these drugs requires a research protocol that can take up to 10 years (or more) before they are approved for human use. In this exercise, you will look at a few representatives of compounds found in your local drugstore or pharmacy.

I. Aspirin

The history of aspirin development is fairly lengthy. Early compounds contained salicylic acid, which caused many people stomach distress. The discomfort was primarily due to the blood loss into the stomach as the compound passed through the stomach lining. Typically, the blood loss is approximately 2 mL, but for many people, it is significantly more. Current aspirin compounds have been slightly modified from earlier compounds by adding an acetyl group to the structure. In an acidic solution, the aspirin is not ionized and it is nonpolar. This allows it to pass through the stomach lining without causing bleeding. However, in a neutral environment, it ionizes and cannot pass back through.

Aspirin (in stomach)

Aspirin (after passing through stomach)

a. What is the molar mass of aspirin? _____

b. What is the charge on the ionized version? _____

c. The half-life of aspirin in the body is approximately 15 minutes. This means that if you take a dose of aspirin, half of it will remain effective in 15 minutes and one-quarter of the original dose will be effective in 30 minutes, etc.... Dosage is determined by body mass, and for aspirin, the dose is 20 mg/kg of body weight per day.

Calculate the amount of aspirin that would remain effective one hour after taking two extra-strength aspirin tablets (500 mg each) if your body mass is 70 kg.

II. Amoxicillin

Amoxicillin is the most prescribed antibiotic in the United States. Primarily, antibiotics are chemical compounds produced by microorganisms that prevent bacterial growth. The dosage is commonly 5.0 mg/kg of body weight, and its half-life in the body is one hour.

Amoxycillin

a. What is the molar mass of amoxicillin? _____

b. A young child is diagnosed with an ear infection, and it is determined that amoxicillin is the best choice of medication. How much amoxicillin should this 22-kg child receive, and how often should it be delivered so that there is no less than 10% of effective antibiotic at work in her system?

III. Research, using references such as the World Wide Web, a drug or medicine that has been recently introduced. Explain the chemistry behind how this medicine works.

IV. Vitamins and Minerals

Complete the following chart about common vitamins and minerals. The web site contains information in Table I that will be helpful. Consider a vitamin and mineral supplement tablet that weighs 0.95 g in your calculations on minerals.

Vitamin/Mineral	Recommended daily intake value including appropriate units	If a vitamin, is it polar or nonpolar? Based on this, what is the potential for overdose? Why?	If the tablet contains 100% of the daily intake, what percent by mass is due to this nutrient?*
Vitamin A (retinol)	*1 RE ≈ 1 µg		
Vitamin C (ascorbic acid)			
Vitamin B$_1$ (thiamin)			
Calcium			*Calcium is not usually taken in 100% dose. (Assume 33% of RDA per tablet.)
Iron			
Zinc			

STUDENT COMPANION FOR CHEMISTRY

Chapter 5

The Properties of Gases

Activity 5.1 Gas Laws—Guided Reading, p. 71
Activity 5.2 Describing and Predicting Changes, Using Gas Law Concepts—Group Problem, p. 75
Activity 5.3 Gas Stoichiometry and the Ideal Gas Law—Group Problem, p. 77
Activity 5.4 Group Challenge Problem, p. 79
Activity 5.5 Graphing Gases: Real and Ideal, p. 81

ACTIVITY 5.1
Gas Laws
Guided Reading

Name _____

Group _____

Read the introduction to Chapter 5 and Section 5.1.

1. List some common gases. (See Appendix 2E for some examples.) For each gas that you list, answer the following questions.

    a. What is the name and chemical formula for this gas?

    b. Where is this gas found and what is it used for?

2. Describe the differences between a gas and the other two states of matter.

STUDENT COMPANION FOR CHEMISTRY

3. What is the kinetic model? How does it describe gas behavior? (See Section 5.13.)

Read the two sections describing pressure, Sections 5.2 and 5.3, and answer the following questions.

4. Define pressure.

5. How could you measure atmospheric pressure? the pressure of a gas sample?

6. What is the SI unit of pressure? What are some other common pressure units used in chemistry? List from smallest to largest pressure unit.

7. Define pressure, using the kinetic model for gases. (See Section 5.13.)

Read Section 5.4 and the beginning of Section 5.5.

8. How does the kinetic model define the temperature of a gas? (See Section 5.13.)

9. What is meant by absolute zero?

10. What are common units for temperature? What is the SI unit of temperature?

11. Convert:
   a. 1.15 atm to Torr

   $1.15 \times 760 = 874$ torr

   b. 895 mmHg to Pa

   $\dfrac{895 \text{ mmHg} \times 1.0325 \times 10^5 \text{ Pa}}{760 \text{ mm Hg}} = 1.19 \times 10^5 \text{ Pa}$

   c. 4 K to °C

   $4 - 273 = -269$

   d. 25°C to kelvins

   $25 + 273.15 = 298.15 \text{ K}$

   e. 3.77 kPa to Torr

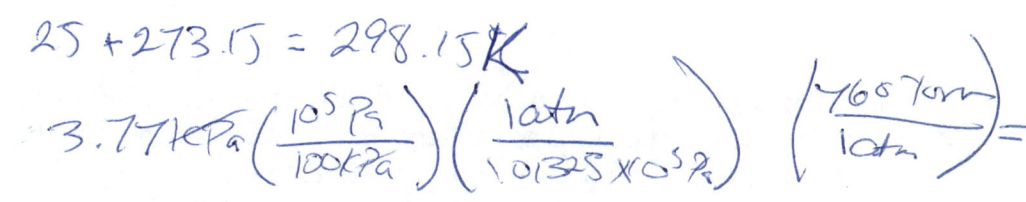

   $3.77 \text{ kPa} \left(\dfrac{10^5 \text{ Pa}}{100 \text{ kPa}} \right) \left(\dfrac{1 \text{ atm}}{1.0325 \times 10^5 \text{ Pa}} \right) \left(\dfrac{760 \text{ Torr}}{1 \text{ atm}} \right) =$

   f. 325.1°C to kelvins

STUDENT COMPANION FOR CHEMISTRY 73

Read the textbook sections on gas laws, Sections 5.4 through 5.8.

12. Write mathematical expressions for the following laws. (Be sure to define all symbols that you use and to list any constant conditions.)

    a. Ideal gas law

    b. Boyle's law

    c. Charles's law

    d. Avogadro's principle

Return to class and answer the following questions.

13. Complete the following plots for an ideal gas. Be sure to label the axes.

    a. V versus T; P and n constant.

    b. P versus V; T and n constant.

    c. V versus n; T and P constant.

    d. P versus T; V and n constant.

14. Develop a strategy sheet for solving ideal gas problems. Be sure to include problems with changing conditions as well as those that do not involve changing conditions. (See Toolbox 5.1 and Section 5.9 of your textbook.)

ACTIVITY 5.2
Describing and Predicting Changes Using Gas Law Concepts
Group Problem

Name _____

Group _____

For each, draw a sketch of the system before and after the change, devise a strategy to solve the problem, and solve the problem.

1. A sample of neon gas that occupies a volume of 533 cm³ at 714 Torr is compressed at constant temperature to 2.37 atm. What is the final gas volume?

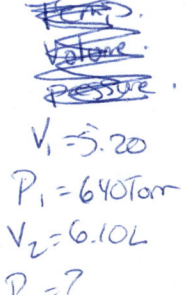

    .93 atm

    $P_2 V_2 = P_1 V_1$

2. A given sample of gas has a volume of 5.20 L at 27°C and 640 Torr. The gas is then heated to 100°C and allowed to expand to 6.10 L. Calculate the new pressure after the expansion.

    $V_1 = 5.20$
    $P_1 = 640 \text{ Torr}$
    $V_2 = 6.10 \text{ L}$
    $P_2 = ?$

    $PV = t$

3. A group of chemistry students has added a 12.5-g sample of a gas to an evacuated, constant volume container at 22°C. The pressure of the gas is greater than atmospheric pressure. Now the students will heat the gas at constant pressure. The easiest way to do this is to allow some of the gas to escape during the heating. What mass of gas must be released if the temperature is raised to 202°C?

ACTIVITY 5.3

Gas Stoichiometry and the Ideal Gas Law

Group Problem

Name _____

Group _____

Consider Exercise 5.48 in your textbook: Nitroglycerin is a shock-sensitive liquid that detonates by the reaction

$$4\ C_3H_5(NO_3)_3(l) \rightarrow 6\ N_2(g) + 10\ H_2O(g) + 12\ CO_2(g) + O_2(g).$$

1. Consult with your group members to develop a strategy for calculating the *total* volume of product gases at a given temperature and pressure from a known amount of nitroglycerin. Record your strategy below.

2. Assign each student in your group one of the following sets of data. Use this data along with your group strategy to calculate the *total* volume of the product gases.

Student	Mass of nitroglycerin	Temperature	Pressure	Total volume of product gases in liters
1	1.0 g	100.0°C	150 kPa	
2	2.0 g	298 K	1.00 atm	
3	1.0 kg	273 K	1.00 atm	
4	250 g	80.0°C	577 mmHg	

STUDENT COMPANION FOR CHEMISTRY

ACTIVITY 5.4
Group Challenge Problem

Name _____

Group _____

Suppose you have 250.0 mL of a solution containing carbonate ions and you want to determine the molar concentration of carbonate ions in this solution. You decide to do this by adding some 1.00 m HCl and capturing the carbon dioxide gas formed by water displacement. After bubbles stop forming in your solution, you assume that all the carbonate ions have reacted with the HCl. You then measure the temperature, pressure, and volume of your gas sample and find $T = 22.0°C$, $P = 756.0$ Torr, and $V = 851$ mL.

1. Write a balanced molecular and net ionic equation for the reaction of hydrochloric acid with a soluble carbonate salt. Write the net ionic equation for this reaction.

2. Calculate the number of moles of CO_2 gas formed if the vapor pressure of water at 22.0°C is 21 Torr.

3. Calculate the molar concentration of carbonate ions in your original solution.

STUDENT COMPANION FOR CHEMISTRY

Now you want to test your solution to see if there are any carbonate ions remaining (maybe you did not add enough HCl). To do this, you add 1.00 M copper(II) chloride (aq).

4. What will happen if carbonate ions remain in your solution?

5. Write a net ionic equation representing what will happen.

6. Suppose 0.618 g of a precipitate form when you add excess aqueous $CuCl_2$. What can you conclude about the concentration of carbonate ions that you calculated in 3? Was it correct, too low, or too high?

7. Calculate a "corrected" molarity of carbonate ions in your original solution, using the additional information provided in 6.

ACTIVITY 5.5
Graphing Gases: Real and Ideal

Name _____

Group _____

For this activity, you will use a graphing tool such as the curve fitter found on the Problem-Solving Skills CD-ROM. Attach a printout of your graph to this sheet.

1. What is the equation for Boyle's law? _____

2. What properties must remain constant when you use this equation?
   _____

3. Use the curve-fitter tool to develop a graph that has P × V on the y-axis and P on the x-axis. Assume that the temperature remains consistent for each gas. Print out a copy of the graph using the data below.

   Pressure of H_2 (at 0°C, in atm)   Pressure of CO_2 (60°C in atm)

   Volume of H_2 (at 0°C, in atm)   Volume of CO_2 (60°C in L)

P, V	P, V
20.0, 5.10	20.0, 4.70
80.0, 1.31	60.0, 1.25
160.0, 0.69	160.0, 0.25
220.0, 0.51	220.0, 0.26
240.0, 0.48	240.0, 0.27

   Sketch on your graph the curve that would be expected for an ideal gas based on your understanding of the ideal gas law under these circumstances.

   Compare any deviations of your graphs for hydrogen and carbon dioxide gases, with the expected results of ideal gases.

STUDENT COMPANION FOR CHEMISTRY

Chapter 6

Thermochemistry: The Fire Within

Activity 6.1	Counting Calories—Group Problem, p. 85
Activity 6.2	Enthalpy—Guided Reading, p. 87
Activity 6.3	Hot Packs and Cold Packs—Demonstration, p. 91
Activity 6.4	Thermite Reaction and Chemical Reactions of Solids—CD-ROM Demonstration, p. 95
Activity 6.5	Calculating the Enthalpies of a Combustion Reaction—Worksheet, p. 99
Activity 6.6	Group Challenge Problem, p. 101

ACTIVITY 6.1
Counting Calories
Group Problem

Name _____

Group _____

Your signature indicates your active participation in solving this group problem.

Your group has been hired to be a consulting team to a local cracker manufacturer. A newly developed product, Mini-munchers, has been advertised to be 40% lower in calories than its counterpart, Mega-munchers. Several difficulties have been discovered in the quality of research for Mini-munchers. The company would like you to evaluate the validity of its advertised claims.

A calorimeter was used to determine the amount of calories in each product. The calorimeter technician used a new constant-pressure (bomb) calorimeter to determine the calories each serving of cracker contained. However, the new calorimeter came from a company called "S.I. Equipment" and the materials allowed for measurements to be taken only in accepted SI.

1. Provide a labeled diagram of what a bomb calorimeter might look like.

One of the first things your team decides to do is to inspect the calorimeter used. Identify and list any factors that prevent this calorimeter, or any calorimeter, from being completely accurate.

2. Oxygen is added and the crackers are oxidized completely. A copy of the technician's notes are included below. Use this information to determine whether Mini-munchers can legally claim to have 40% fewer calories than the same serving size of Mega-munchers.

Mini-munchers	*Mega-munchers*
500.0 mL water used in calorimeter	500.0 mL water used in calorimeter
Beginning temp = 298 K	Beginning temp = 307 K
Final temp = 322 K	Final temp = 340 K

The sample size used was 0.100 serving.

Water was used in the calorimeter and the mass of the cracker was negligible, so the specific heat that was used in the calculations was based on water alone (1 cal/g°C). One tenth of a serving-sized portion, as listed on the package, was used for each test.

Food products typically list kilocalories on the label, called food Calories, rather than calories to ease the minds of consumers.

How many kcal were found in a serving of Mega-munchers? _____

How many kcal were found in a serving of Mini-munchers? _____

To impress the executives at the meeting when you present your results, you decide to include joule values in your calculations as well.

How many joules were found in a serving of Mega-munchers? _____

How many joules were found in a serving of Mini-munchers? _____

On the basis of your calculations, write a brief report to the company, indicating your conclusion about whether the company can legally claim that the Mini-muncher product contains 40% fewer calories than Mega-munchers. Justify your conclusion.

ACTIVITY 6.2
Enthalpy
Guided Reading

Name _____

Group _____

Read Sections 6.9 through 6.11 in your textbook.

Answer the following questions individually after you have read the required sections about enthalpy.

1. Explain, in your own words, what **enthalpy** is. Include in your explanation any symbols and units that are used to describe the enthalpy of a reaction.

2. Distinguish between the terms **sublimation** and **vaporization.**

3. Distinguish between the following terms:
   a. melting and dissolving

   b. vaporization and fusion

STUDENT COMPANION FOR CHEMISTRY

4. Refer to Section 6.12 in your text. Explain the relationship that exists between enthalpy and type of chemical process.

5. For each of the following examples, write in the blank an arrow that shows the direction of enthalpy change (↑ = increase and ↓ = decrease) of the system when

   _____ a. a reaction gives off heat

   _____ b. a liquid vaporizes

   _____ c. propane undergoes combustion

   _____ d. carbon dioxide sublimes

   _____ e. water freezes into ice

   _____ f. a beaker becomes colder as a reaction occurs

6. Complete the homework problems assigned by your instructor.

7. Complete the following graph, from Fig. 6.27, before you complete the remainder of these questions with your group.

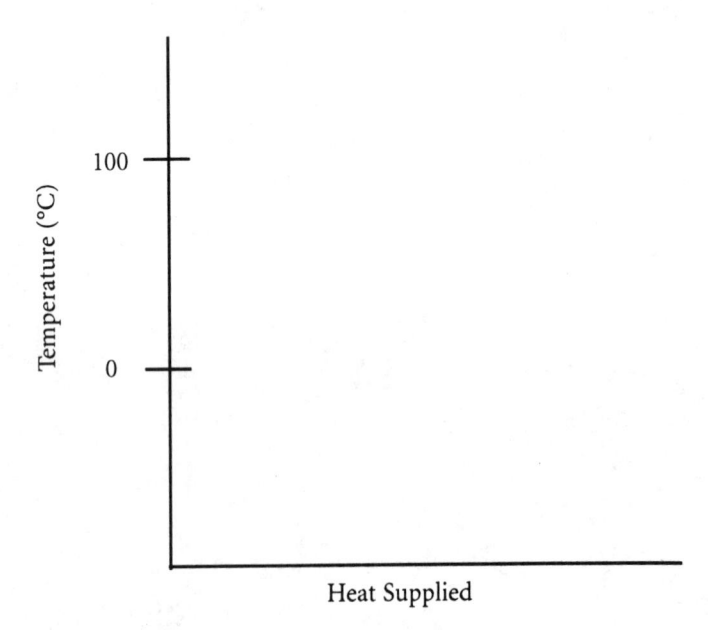

8. a. As a group, make a list of the important pieces of information from your graph that would allow you to solve a problem in which you are asked to calculate the change of enthalpy from one state to another.

b. What information, if any, is needed that is not directly found on this chart?

9. Calculate the change in enthalpy that would occur when heating 3 mol of water from solid at 0°C to vapor at 100°C.

10. Calculate the change in enthalpy that would occur when 90.0 g of water vapor at 100°C is cooled to form ice at 0°C.

ACTIVITY 6.3
Hot Packs and Cold Packs

Name _____

Group _____

Your signature above indicates your active participation in this group exercise.

1. Your instructor will demonstrate the operation of a hot pack in which the heat is produced by a chemical reaction. Although there may be many substances present that are not involved in the reaction, identify

   a. the reactants

   b. the products

   c. Write a balanced equation for this reaction.

   d. What is the purpose of the addition of the components that are not involved directly with this reaction?

STUDENT COMPANION FOR CHEMISTRY

e. Diagram or discuss what combining the reactants might look like from a molecular level.

f. Determine the enthalpy change associated with this reaction. (Consider whether this value should be positive or negative and why.) Assume there are 5.0 g of iron in the handwarmer.

g. Without calculating, what would you expect to happen to the enthalpy of this reaction if you were to double the amount of all the reactants? Explain your reasoning.

2. Your instructor will demonstrate a different reaction or assign one from your CD-ROM. Identify

   a. the reactants

   b. the products

c. Write a balanced chemical equation for this reaction.

d. Diagram or discuss what adding the reactants together might look like from a molecular standpoint.

e. Determine the enthalpy change associated with this reaction. Assume that you are starting with 50.0 g of the ammonium compound and excess amounts of any other reactant materials.

3. Look at the construction of a commercially prepared hot pack found in drug or sporting goods stores. What precautions have been taken to assure consumer safety?

ACTIVITY 6.4

CD-Demonstration
Viewing Guide

Name _____

Group _____

View the demonstrations titled "Thermite Reaction" and "An Endothermic Reaction of Solids" on the Problem-Solving Skills CD-ROM.

Thermite Reaction

1. Write a balanced equation for the thermite reaction.

2. Is this reaction endothermic or exothermic? How can you tell?

3. This is an example of an oxidation-reduction reaction. What is oxidized and what is reduced in this reaction?

4. Explain the two separate segments of the thermite reaction according to the demonstration explanation on the CD.

STUDENT COMPANION FOR CHEMISTRY

5. Predict the minimum temperature that exists when product is formed in the thermite reaction. Explain your reasoning.

6. Discuss a practical use for this reaction.

Chemical Reactions of Solids

7. Write the reaction that occurs when barium hydroxide is reacted with ammonium thiocyanate.

8. What type of reaction is illustrated by the preceding reaction: exothermic or endothermic?

9. Is the rate of this reaction faster or slower than the thermite reaction? Propose an explanation for your answer.

10. Discuss a practical use for this type of a reaction.

ACTIVITY 6.5
Calculating the Enthalpy of a Reaction
Worksheet

Name _____

Group _____

---

1. A thermochemical equation contains both the balanced chemical equation and the enthalpy change that accompanies the reaction.

   a. Write a balanced chemical equation for the combustion of ethene (C_2H_4) in excess oxygen.

   b. Explain your strategy for determining the total enthalpy change from the enthalpies of formation for all reactants and products.

   c. Use the following table to calculate the *total* enthalpy change in the reaction. (Remember that enthalpy is a state function, which means that a change in enthalpy is determined by the initial and final states of the system.)

Standard enthalpies of formation in (kJ/mol)	
CH_4	−74.8
C_2H_4	52.3
C_2H_6	−84.7
CO	−110.5
CO_2	−393.5
$H_2O(l)$	−285.8

STUDENT COMPANION FOR CHEMISTRY

d. Explain why the enthalpy of formation of an element in its standard state is 0.

e. Write a thermochemical equation that describes what happens when ethane (C_2H_6) undergoes combustion in the presence of excess oxygen.

f. Explain why there is a difference in the reaction enthalpies between the combustion of ethene and the combustion of ethane.

2. A 64.01-g sample of methane (CH_4) is combined with excess oxygen and burned in a calorimeter. Write a thermochemical equation for this reaction.

3. What is enthalpy density and what role does it play when rating fuels? What other factors need to be taken into consideration when comparing and rating fuels?

ACTIVITY 6.6
Challenge Problem

Name _____

Group _____

Many vehicles on the road are equipped with air bags for safety. These bags are designed to inflate in moderate to severe frontal or near-frontal crashes. The air bag inflates if the velocity of the impact is more than approximately 22 km/hr (16 mph). Some vehicles have a higher impact velocity minimum, but this value is the average for many vehicles.

If the frontal impact is severe enough, the air bag sensing systems detect that the vehicle is stopping because of a crash. An electrical impulse triggers a chemical "fuse," which releases heat energy. The air bag itself is filled with sodium azide (NaN_3), which then undergoes a decomposition reaction to produce the nitrogen gas that fills the sealed cloth bag. Sodium metal is also formed in this reaction.

1. Write a chemical equation for the decomposition of sodium azide.

The inflation of the air bag takes less than 0.05 s. Deflation quickly follows. Air bags are designed to open once. Owner's manuals recommend that you take your car back to the dealer to have the system reinstalled after use. You have been hired by the manufacturer to make sure that systems are properly packaged for use. Therefore, you need to find out the following information.

2. What mass of sodium azide is needed to inflate an air bag to 50.0 L at 298 K and 1.00 atm pressure?

3. a. How many grams of sodium metal will be produced when an air bag is inflated to 50.0 L?

   b. What problems are associated with producing sodium metal? What would happen if the air bag broke open and sodium got on the car driver or on a passenger?

STUDENT COMPANION FOR CHEMISTRY

c. As a result of the ignition of sodium nitrate in the fuse, heat is produced. Calculate the amount of heat this reaction produces.

4. Calculate the amount of heat generated when sodium azide in an air bag decomposes (ΔH_f° for sodium azide at 25°C is 21.7 kJ/mol).

Chapter 7

Inside the Atom

Activity 7.1	Spectroscopy—Hands-On Group Activity, p. 105
Activity 7.2	Quantum Mechanical Model—Guided Reading, p. 107
Activity 7.3	Quantum Mechanical Model Concept Map—Group Problem, p.113
Activity 7.4	Electron Configurations—Worksheet, p. 115
Activity 7.5	Periodic Trends—Group Problem, p. 119

ACTIVITY 7.1
Spectroscopy
Hands-On Group Activity

Name _____

Group _____

Remove your diffraction grating and look through it at the light sources provided by your instructor. Record your group's discussion below.

Light source	Observations: Without grating	Observations: Wih grating

Discuss explanations of your observations, relating them to what you understand about quantum mechanical theory. Record your explanations below.

ACTIVITY 7.2
Quantum Mechanical Model
Guided Reading

Name _____

Group _____

Read Section 7.5 in your textbook and answer the following questions.

1. Which scientist was responsible for the current model used to describe electron motion in atoms?

2. How is the behavior of electrons described according to the quantum mechanical model? How is this similar to the particle and wave descriptions of electromagnetic radiation?

3. How is the location of an electron described?

4. Define atomic orbital according to the quantum mechanical model.

5. a. What four letters are commonly used to identify the shapes of orbitals?

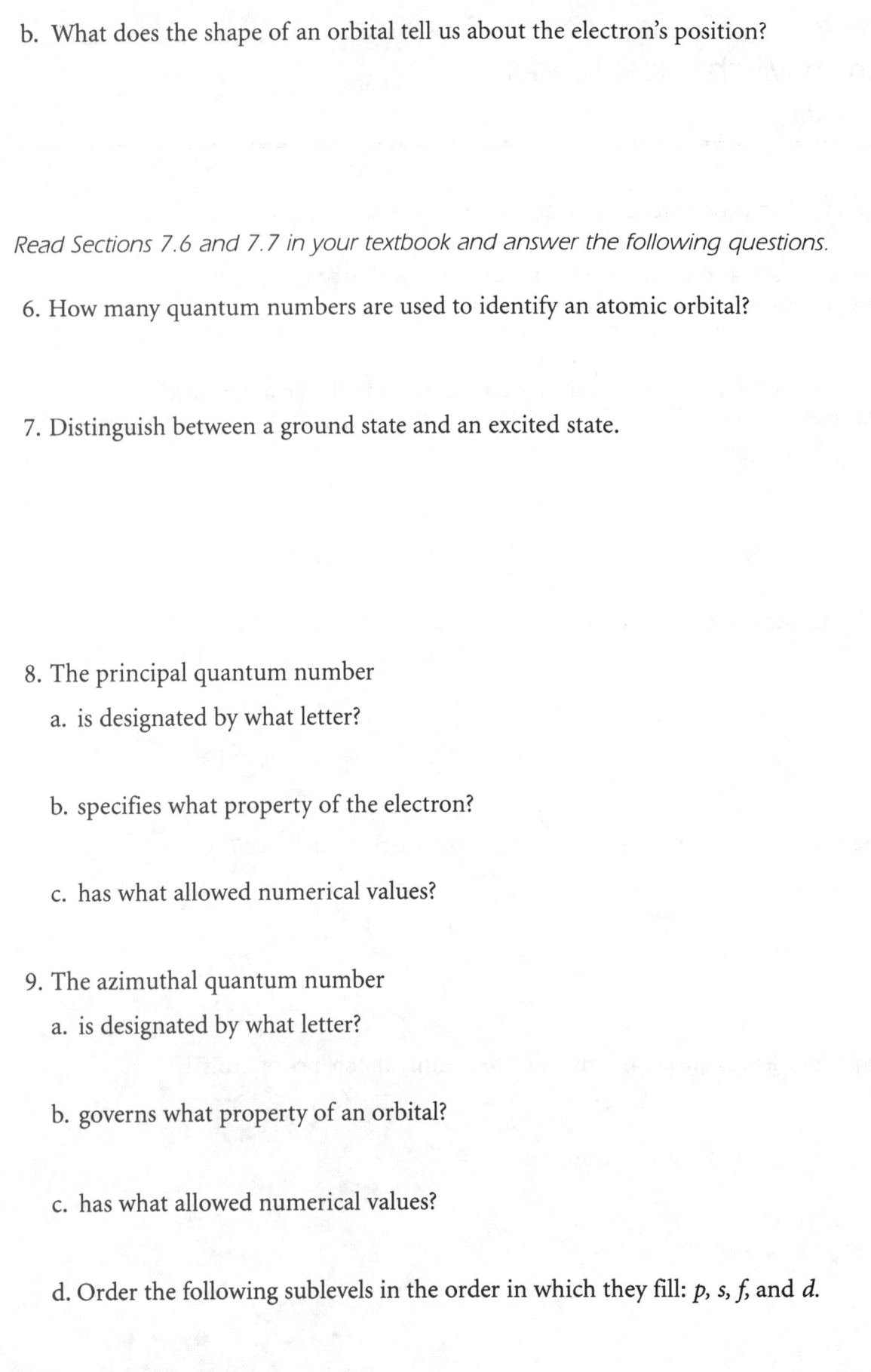

b. What does the shape of an orbital tell us about the electron's position?

Read Sections 7.6 and 7.7 in your textbook and answer the following questions.

6. How many quantum numbers are used to identify an atomic orbital?

7. Distinguish between a ground state and an excited state.

8. The principal quantum number
    a. is designated by what letter?

    b. specifies what property of the electron?

    c. has what allowed numerical values?

9. The azimuthal quantum number
    a. is designated by what letter?

    b. governs what property of an orbital?

    c. has what allowed numerical values?

    d. Order the following sublevels in the order in which they fill: *p, s, f,* and *d.*

e. Fill in the following table.

n	Number of subshells	Letter designations	Numerical values
1			
2			
3			
4			

f. Describe or sketch the shapes of an *s* and a *p* orbital.

10. The magnetic quantum number

   a. is designated by what letter?

   b. labels what property of the orbitals?

   c. has what allowed values?

   d. Fill in the following table.

Subshell	Number of orbitals	Maximum number of electrons
s		
p		
d		
f		

Read Section 7.8 and answer the following questions.

11. What is the name of the quantum number used to describe the spin of an electron inside an atom?

12. What letter is used to represent this quantum number? What property of the electron does it describe?

13. What are the allowed values of the spin magnetic quantum number?

Read "Investigating Matter 7.1," which describes how spin states were discovered, and discuss the following review questions with your group members.

14. Which energy level is closer to the nucleus $n = 1$ or $n = 3$? Which is lower in energy?

15. How many subshells are allowed in the energy level with $n = 10$? Is this energy level found in a ground state or an excited state for known elements? What is the highest value of n for ground states of known elements?

16. Sketch or describe the relative shapes of a *1s*-orbital and a *2s*-orbital.

17. The maximum number of electrons possible in any orbital is 2. The maximum number of electrons possible in a principal energy level is obtained by using the following formula: maximum number of electrons = $2n^2$, where n is the principal quantum number. Justify this formula.

18. Work the textbook exercises assigned by your instructor.

ACTIVITY 7.3
Quantum Mechanical Model Concept Map
Group Problem

Name _____

Group _____

Use the space below to generate a concept map for the quantum mechanical model. A concept map should show (using lines and arrows) the relationships between the concepts or ideas related to a particular topic (quantum mechanical model). In the center of the page is written the main topic: quantum mechanical model. Write down as many concepts related to or part of this model as you can and show their connections to one another.

QUANTUM MECHANICAL

MODEL

ACTIVITY 7.4
Electron Configurations
Worksheet

Name _____

Group _____

1. Fill in the orbital diagrams by writing the electron configurations for the following neutral atoms in their ground states. Use ↑ to represent a single electron and ↑↓ to represent a pair of electrons.

	1s	2s		2p	
a. H	↑				
b. He	↑↓				
c. Li	↑↓	↑	☐	☐	☐
d. Be	↑↓	↑↓	☐	☐	☐
e. B	↑↓	↑↓	↑	☐	☐
f. C	↑↓	↑↓	↑	↑	☐
g. N	↑↓	↑↓	↑	↑	↑
h. O	↑↓	↑↓	↑↓	↑	↑
i. F	↑↓	↑↓	↑↓	↑↓	↑
j. Ne	↑↓	↑↓	↑↓	↑↓	↑↓

	3s		3p	
k. Na [Ne]	↑	☐	☐	☐
l. Mg [Ne]	↑↓	☐	☐	☐

	3s	3p

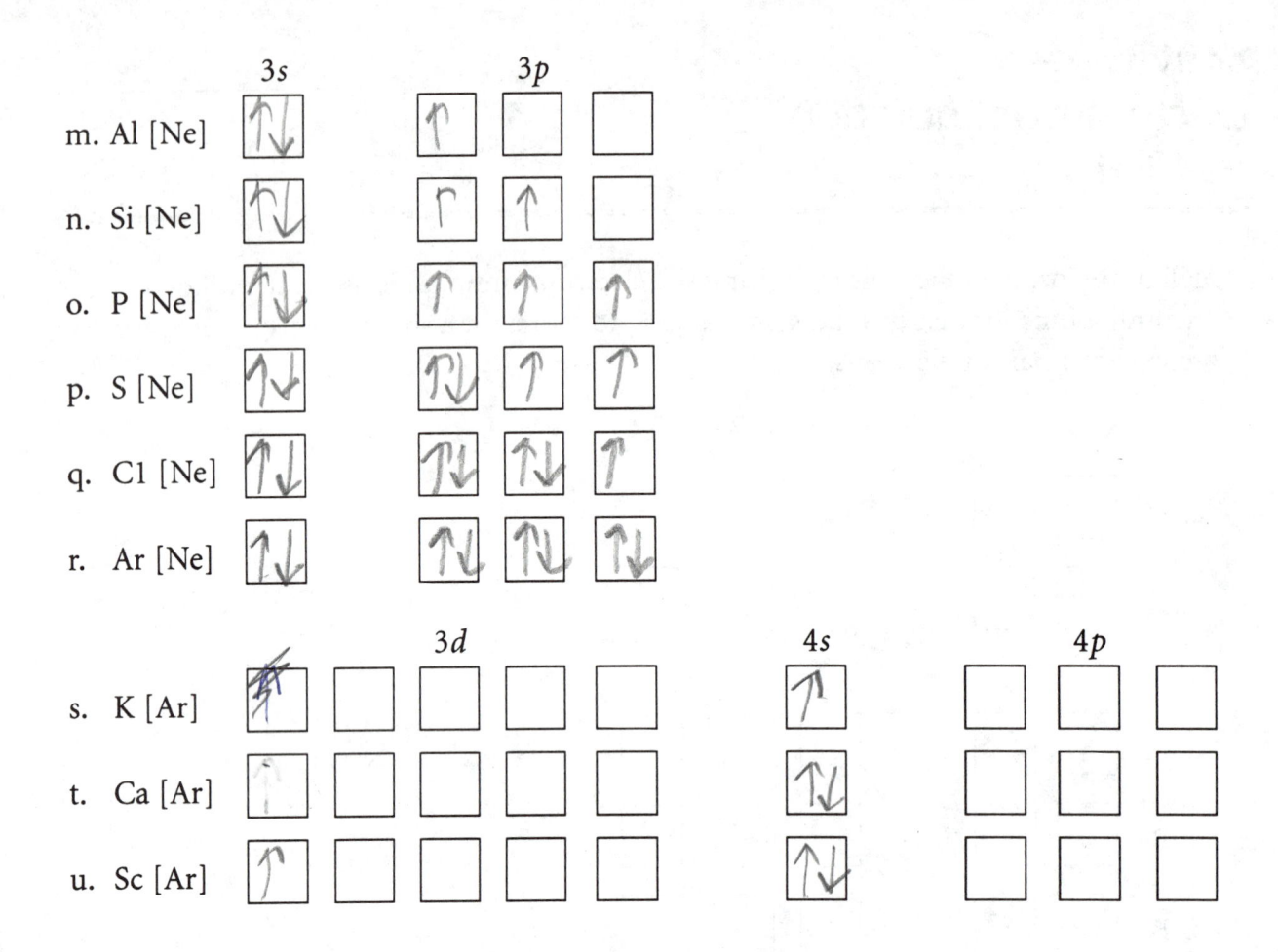

m. Al [Ne] [↑↓] [↑][ ][ ]

n. Si [Ne] [↑↓] [↑][↑][ ]

o. P [Ne] [↑↓] [↑][↑][↑]

p. S [Ne] [↑↓] [↑↓][↑][↑]

q. Cl [Ne] [↑↓] [↑↓][↑↓][↑]

r. Ar [Ne] [↑↓] [↑↓][↑↓][↑↓]

	3d	4s	4p

s. K [Ar] [ ][ ][ ][ ][ ] [↑] [ ][ ][ ]

t. Ca [Ar] [ ][ ][ ][ ][ ] [↑↓] [ ][ ][ ]

u. Sc [Ar] [↑][ ][ ][ ][ ] [↑↓] [ ][ ][ ]

2. Indicate in the blank periodic table below which orbitals are being filled in which regions.

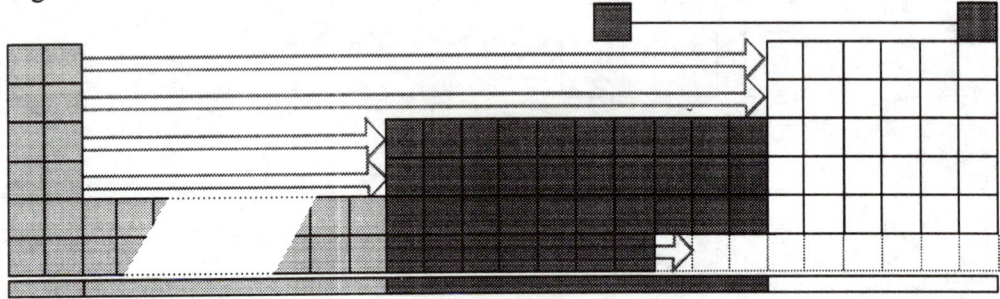

3. Write electron configurations for the following ions:

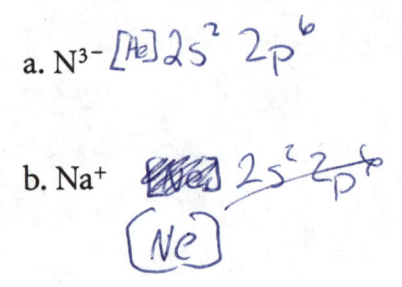

a. N^{3-} [He] $2s^2 2p^6$

b. Na^+ [Ne] $2s^2 2p^6$
   (Ne)

116

CHAPTER 7

c. Cl⁻ ~~[...]~~ [Ar]

d. Mg²⁺ [Ne]

e. Fe²⁺ [~~Cr~~] [Ar]3d⁵

f. Fe³⁺ [~~V~~] [Ar] [

g. Sn²⁺ [Cd]

h. Sn⁴⁺ [Pd]

ACTIVITY 7.5

Periodic Trends

Group Problem 11/13/01

Name ______________________

Group ______________________

As homework read "Investigating Matter 7.2" along with Sections 7.15–7.19.

1. Define effective nuclear charge.

2. Use the diagram of the periodic table given below to indicate the trends in atomic radius. Use arrows and words to indicate an increase or a decrease in size.

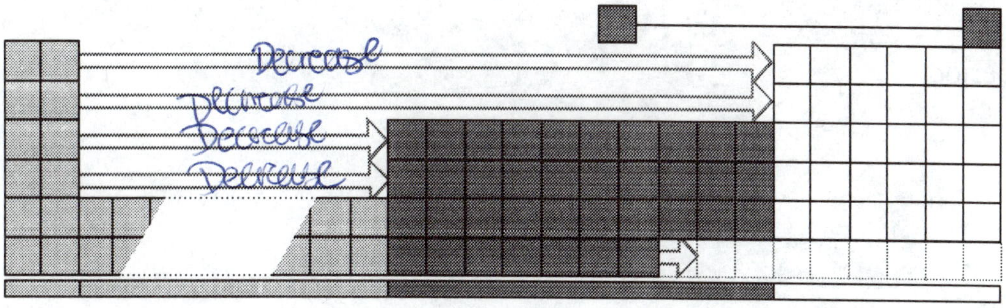

3. Explain the periodic trends for atomic radius.

   a. down a group:

   Increase

   b. across a period:

   Decrease

4. Use the diagram of the periodic table given below to indicate the trends in first ionization energy. Use arrows and words to indicate an increase or a decrease in first ionization energy.

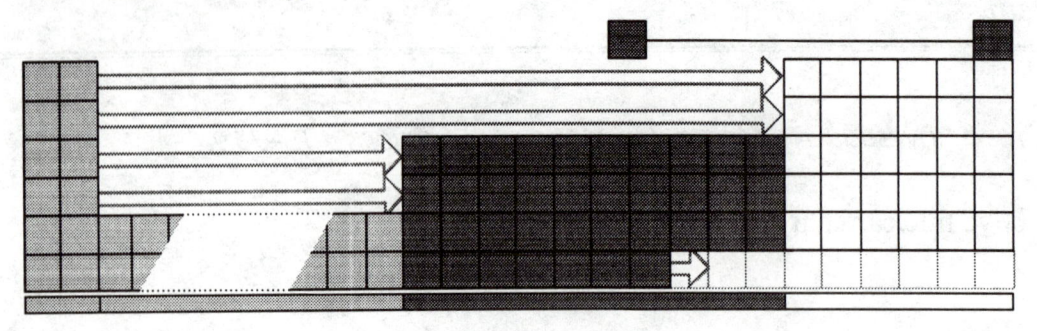

5. Explain the periodic trends for first ionization energy.

   a. down a group:

   b. across a period:

6. Work textbook Exercise 7.97.

7. The most reactive metals are found in the lower left corner of the periodic table, whereas the most reactive nonmetals are found in the upper right. Explain why.

120 CHAPTER 7

Chapter 8

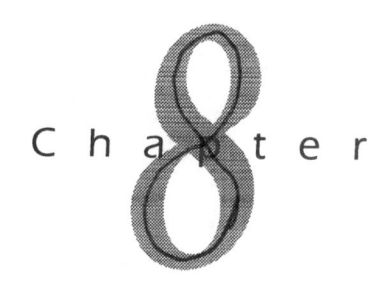

Chemical Bonds

Activity 8.1   Ionic Bonding—Worksheet, p. 123
Activity 8.2   Covalent Bonding—Worksheet, p. 127
Activity 8.3   Ionic versus Covalent Compounds—Group Problem, p. 129
Activity 8.4   Group Modeling of Lewis Dot Structures, p. 131
Activity 8.5   Resonance—Guided Reading, p. 133
Activity 8.6   Review Worksheet, p. 137

ACTIVITY 8.1
Ionic Bonding
Worksheet

Name _____

Group _____

---

1. What happens to valence electrons on two atoms that form an ionic bond?

   The valence electrons are shared.

2. Why are valence electrons the only electrons considered in a Lewis dot structure?

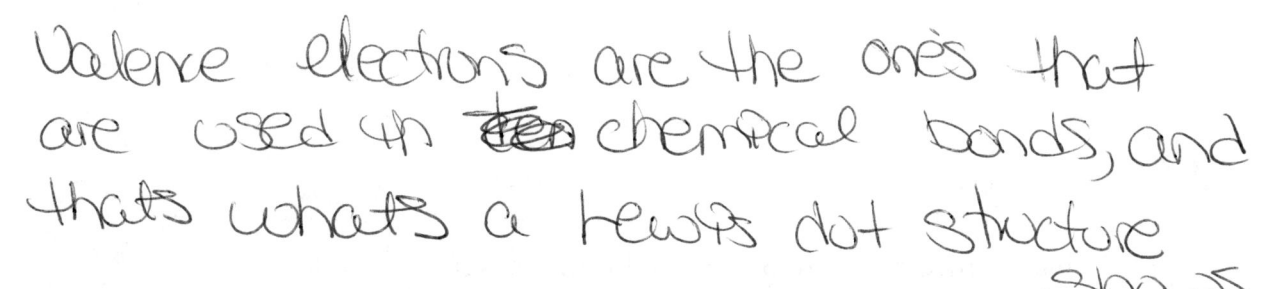

   Valence electrons are the ones that are used in ~~tea~~ chemical bonds, and that's what's a Lewis dot structure shows.

3. Write the Lewis dot structures for the following ionic compounds.

   RbO

   Ca :Ö:

   MgCl$_2$

   Mg^{2+} [:Cl—Cl:]

   KBr

   K—Br:

4. Explain the octet rule, using your own words.

   The octet rule is a transfer of electrons ($s^2 p^6$) configuration)

   — the ideal state to be in the valence shell is full.

STUDENT COMPANION FOR CHEMISTRY   123

5. Are there exceptions to the octet rule? If yes, give an example of an exception. If no, explain why not.

6. Explain how formation of ions lowers energy.

7. Diagram what a lattice structure of sodium bromide, NaBr, would look like.

8. In terms of energy, what happens when an ionic solid vaporizes?

9. Use the Born-Haber process to calculate the lattice enthalpy of $BeCl_2$. The first ionization energy for Be is 1760.0 kJ/mol and the second ionization energy for Be is 2659.S kJ/mol. Use Appendix 2A for additional information.

10. Explain why you do not need to indicate the direction of a lattice enthalpy value.

11. Because of the attraction of opposite charges, ionic solids have some definite characteristics. Describe the following properties of an ionic solid.

its boiling point _____

its melting point _____

how it dissolves in water _____

ACTIVITY 8.2

Covalent Bonding
Worksheet

Name _____

Group _____

---

1. What type of elements form only covalent compounds with one another: metals or nonmetals? Why can't these elements form cations?

2. Before you begin diagramming covalently bonded compounds, you need to determine a few things. Use CH_4, methane, as an example.

   a.

Characteristic	Carbon	Hydrogen
Number of valence e⁻	4	1
Number of atoms present	1	4
Total number of valence e⁻	4 + 4 = 8	
(No. of valence e⁻ needed to form an octet or a duet)	4	4

   b. Divide the total number of available valence electrons by 2:
   Why is this number divided by 2?

   c. Which element is placed in the center of the diagram? Why?

   C

   d. Based on this information, diagram a Lewis dot structure of methane.

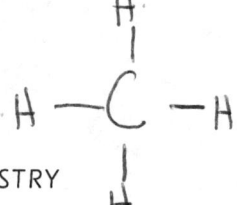

   ```
        H
        |
   H — C — H
        |
        H
   ```

STUDENT COMPANION FOR CHEMISTRY

3. Diagram the Lewis dot structures for the following covalent compounds.

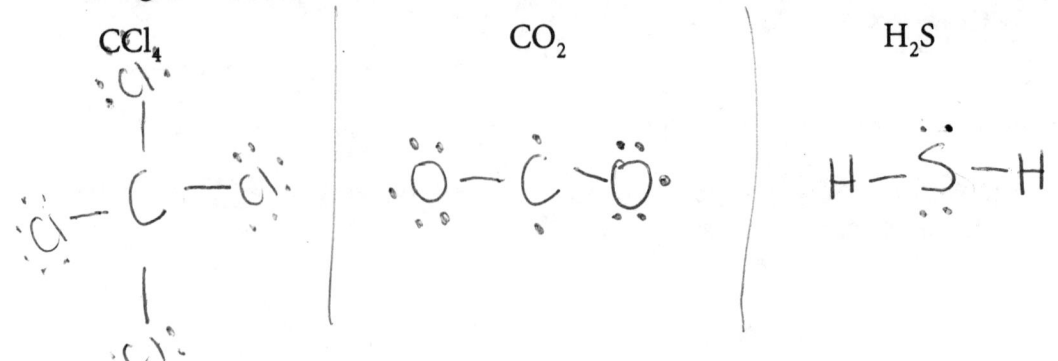

CCl_4

CO_2

H_2S

4. Predict the relative values of the following properties of covalent compounds.

boiling point _____

melting point _____

electrolyte or nonelectrolyte? _____

5. Draw Lewis dot structures of chlorine gas (Cl_2) and nitrogen gas (N_2).

ACTIVITY 8.3
Ionic versus Covalent Compounds
Group Problem

Name _____

Group _____

Your signature at the top of this paper indicates your active participation in this group assignment.

1. a. Develop a group definition (other than the text definition) of electronegativity.

   b. In general, what periodic trends in electronegativity are seen. (You are welcome to sketch a periodic table if you wish, or simply explain with words.)

2. Below is a chart that compares characteristics of ionic and covalent compounds with respect to electronegativity differences. Make a graph from this data with the x-axis representing the electronegativity and the y-axis representing the percentages. Plot separate curves for ionic and covalent character on the same graph.

Difference in electronegativity	0.065	1.19	1.67	2.19
Percentage of covalent characteristics	90	70	50	30
Percentage of ionic characteristics	10	30	50	70

   a. From your graph, estimate the approximate electronegativity difference value that determines whether bonds will be ionic or covalent?

STUDENT COMPANION FOR CHEMISTRY

b. Discuss, in your group, how it is possible for a compound to have both ionic and covalent characteristics. Write down some main points of your discussion.

c. Use a table of electronegativities. Decide which group member will be 1, 2, 3, and 4. Each individual is responsible for identifying and naming a compound whose atoms have the approximate electronegativity difference given. In addition, include a Lewis dot structure of this compound and a brief description of the characteristics that you predict for this compound. **Your information must be accepted by your group before it will be accepted by your instructor.**

Person One is _____ Electronegativity difference: 1.8

Person Two is _____ Electronegativity difference: 1.0

Person Three is _____ Electronegativity difference: 3.0

Person Four is _____ Electronegativity difference: 0.4

ACTIVITY 8.4
Group Models of Lewis Dot Structures

Name _____

Group _____

Determine who will be the reporter, secretary, leader, and model builder in your group. For the molecules listed,

1. draw the correct Lewis structure.
2. build a model for each molecule.
3. circle any molecules that have lone pairs of electrons on the central atom.

Molecules

CS_2 HCN H_2O

SO_3 NO_2^-

CCl_4 NCl_3

STUDENT COMPANION FOR CHEMISTRY   131

ACTIVITY 8.5
Resonance
Guided Reading

Name _____

Group _____

Read Section 8.8 in your textbook and answer the following questions about resonance structures.

1. Explain the term **resonance** as it applies to bonding.

2. a. Which of the following bonds would you expect to be the longest between the same two elements: a single bond, a double bond, or a triple bond? Explain your reasoning.

   b. Based on your explanation for 2a, which bond would be the shortest?

3. Which type of bond (single, double, or triple) would you expect to provide the most energy? Explain your reasoning.

STUDENT COMPANION FOR CHEMISTRY

4. What information would indicate to you that a structure could be a resonance structure?

5. Cite a reason that would explain why a molecule would exhibit resonance rather than one specific bonding pattern.

6. Complete the homework problems assigned by your instructor.

Complete the following questions with your group.

7. Diagram the Lewis dot structure for the acetate ion ($CH_3CO_2^-$). Does this ion exhibit resonance? If yes, how can you tell? If no, why not?

8. a. Look at the table below. Sketch a bar graph from the data listed with each bar representing a bond type. You may want to use two different colors to represent the two types of bonds.

LENGTH OF BOND (pm)

Atoms bonded	Single bond	Double bond	Triple bond
Carbon-carbon	154	134	120
Carbon-nitrogen	143	138	116

b. Benzene is an organic molecule (C_6H_6) that exhibits C—C bond lengths of approximately 140 pm. Explain this, using the information in your graph.

c. Diagram the structure(s) of benzene.

STUDENT COMPANION FOR CHEMISTRY

ACTIVITY 8.6
Review Worksheet

Name _____

Group _____

Name the chemical substance.	Diagram the Lewis dot structure.	If it has resonance structures, diagram another.	Calculate the formal charges on the central atom.	What type of bond is present?	Is it a Lewis acid, Lewis base, or neither?
SO_4^{2-} _____					
$COCl_2$ _____					
CO_3^{2-} _____					
MgF_2 _____					
BCl_3 _____					

STUDENT COMPANION FOR CHEMISTRY

Chapter 9

Molecules: Shape, Size, and Bond Strength

Activity 9.1   Gumdrop Molecular Architecture—Hands-On Group Activity, p. 141
Activity 9.2   Molecular Shapes and Polar Molecules—Group Problem, p. 143
Activity 9.3   Bond Strength—Worksheet, p. 145
Activity 9.4   Orbitals and Bonding—Guided Reading, p. 147
Activity 9.5   Group Challenge Problem, p. 151

ACTIVITY 9.1
Gumdrop Molecular Architecture
Hands-On Group Activity

Name _____

Group _____

In this activity, you will use gumdrops and toothpicks to explore the shapes of molecules.

For these "generic" molecules, design a three-dimensional structure that arranges the valance shell electron pairs about the central atom so that the electron pairs are kept as far away from one another as possible, thus minimizing the electron-pair repulsions. Assume that all the valence electrons are used to form bonds; there are no nonbonding or lone pairs around the central atom. A = central atom; X = atom attached to A. Draw each structure, and in your sketch indicate the X-A-X bond angles. Use the gumdrops, toothpicks, and protractor to help you design each generic molecule.

Hints: Create a first try and look at the bond angles. Is there a small angle that could be made larger? Remember, you want the best overall separation of the bonded atoms. Start with AX_2, and then add another atom. Think how the other bonds are being affected by the new one and move them away from the new bond to compensate. Continue this process with each new atom.

AX_2

AX_3

AX_4

AX_5

AX_6

STUDENT COMPANION FOR CHEMISTRY

ACTIVITY 9.2
Molecular Shapes and Polar Molecules
Group Problem

Name _____

Group _____

Homework

1. Draw a Lewis dot structure and predict the electronic and molecular geometries for the following molecules. Sketch the molecular shape for each molecule and show partial charges for all polar bonds.

    a. CO_2

    b. HCN

    c. BF_3

    d. CH_2O (carbon is the central atom)

    e. SO_2

    f. CH_4

STUDENT COMPANION FOR CHEMISTRY

g. CH$_3$Cl

h. CH$_2$Cl$_2$

i. NH$_3$

j. H$_2$O

Class Activity

2. Exchange papers with group members and correct the Lewis structures that you drew in Question 1. After all members of your group agree on the shapes for the above molecules, answer and discuss the following questions.

   a. Which of the molecules in Question 1 are polar molecules? (*Hint:* Build models to help you decide.)

   b. Use your homework and your discussion of Question 2a to generate a table relating molecular shapes to whether or not a molecule is polar.

ACTIVITY 9.3
Bond Strength
Worksheet

Name _____

Group _____

1. Estimate the enthalpy change that occurs when each of the following molecules is dissociated into its gaseous atoms. Use the data in Table 9.3.

    a. ethane: CH_3CH_3

    b. ethene: $CH_2=CH_2$

    c. benzene: C_6H_6

    d. ethyne: $CH\equiv CH$

2. a. Which molecule in the above list has the longest C—C bond?

    b. Which has the shortest C—C bond?

    c. Which has the strongest C—C bond?

    d. Which has the weakest C—C bond?

3. Benzene has a C—C bond enthalpy that is between the C—C single and C—C double bond enthalpies. Explain this observation, using the information in Table 9.4 and Lewis structures.

STUDENT COMPANION FOR CHEMISTRY

4. Use the information in Tables 9.2 and 9.3 to estimate the reaction enthalpy for the combustion of one mole of each of the compounds listed below. (*Hint:* Write a balanced equation first.) Compare your answers with the enthalpies of combustion given in Appendix 2A of your textbook and explain any differences.

   a. CH_4

   b. C_6H_6

ACTIVITY 9.4
Orbitals and Bonding
Guided Reading

Name _____

Group _____

Read the introductory section (Valance Bond Theory) and Section 9.9.

1. a. Define and sketch a picture of a σ-bond.

   b. What types of atomic orbitals can overlap to form a σ-bond?

2. a. Define and sketch a picture of a π-bond.

   b. What types of atomic orbitals can overlap to form a π-bond?

STUDENT COMPANION FOR CHEMISTRY

3. Make a table showing how many σ-bonds and π-bonds are found in a single, double, and triple bond.

Read the textbook sections on hybridization: Sections 9.10, 9.11, and 9.12.

4. What is a hybrid orbital?

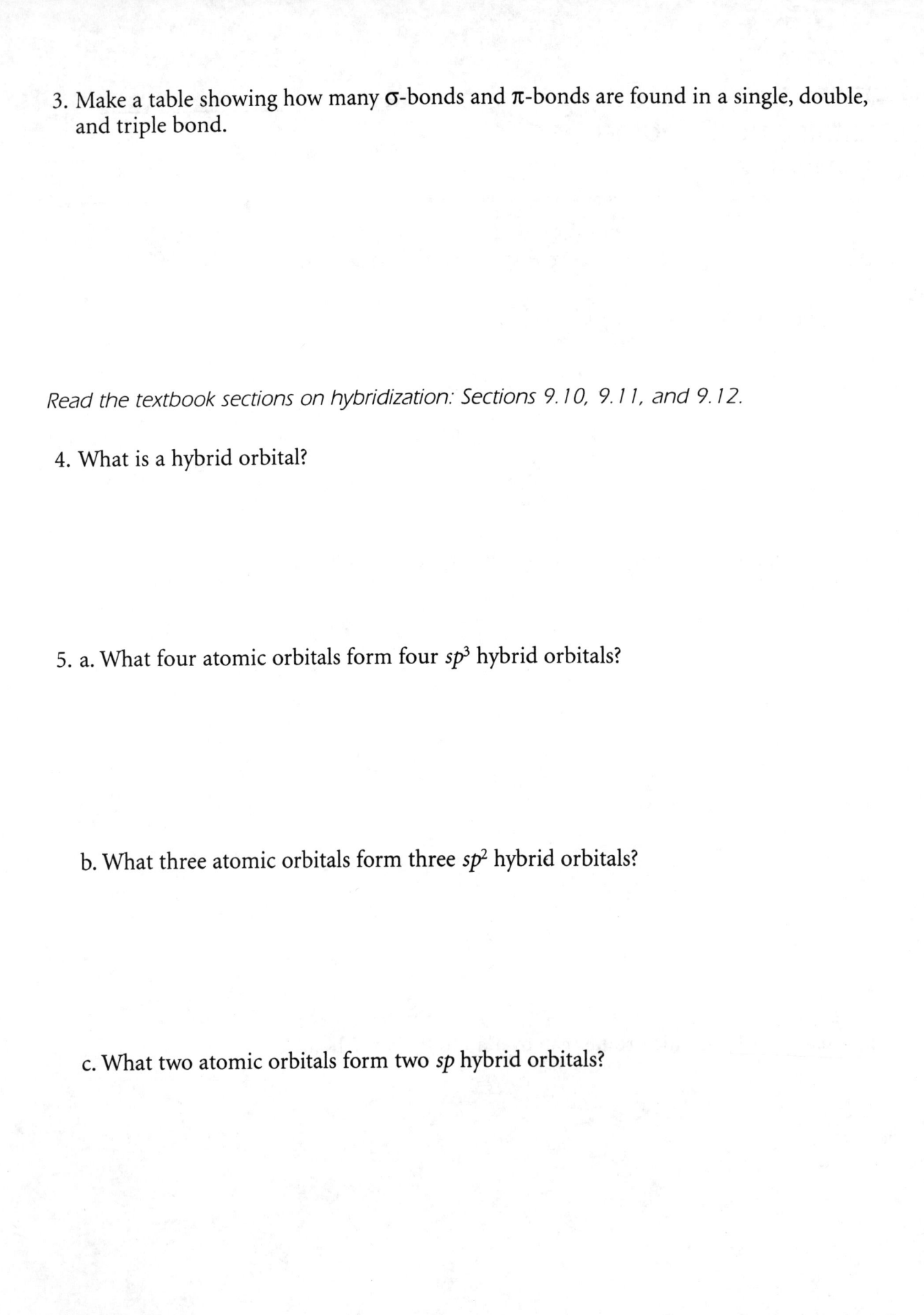

5. a. What four atomic orbitals form four *sp³* hybrid orbitals?

   b. What three atomic orbitals form three *sp²* hybrid orbitals?

   c. What two atomic orbitals form two *sp* hybrid orbitals?

d. What five atomic orbitals form five *dsp³* hybrid orbitals?

e. What six atomic orbitals form six *d²sp³* hybrid orbitals?

6. Fill in the following table, listing each of the five types of hybridization.

Hybridization type	Molecular shape name	Example

Return to class and work the following exercises with your group members.

7. Do lone pairs need to be considered when determining the hybrid orbital type used by an atom?

8. Draw the Lewis dot structures for each of the following molecules. Determine the hybridization type for each central atom, label bond angles, and count the number of σ-bonds and π-bonds for each molecule listed below.

   a. acetic acid, CH_3COOH

STUDENT COMPANION FOR CHEMISTRY

b. CO₂

c. HCN

d. XeF₄

ACTIVITY 9.5
Molecules: Shape, Size, and Strength
Group Challenge Problem

Name _____

Group _____

A chemist is studying two fuels:

Fuel 1 is ethane, C_2H_6. Determine the empirical and molecular formulas for fuel 2 from the following data. Fuel 2 contains only carbon and hydrogen. Combustion analysis of a sample of fuel 2 produces 4.40 g of carbon dioxide and 0.90 g of water. A separate experiment found that the molar mass of fuel 2 is 26 g/mol.

2. Write a balanced chemical equation for the combustion of

   a. fuel 1 (ethane)

   b. fuel 2

3. Calculate the molar enthalpy of combustion for reactions 2a and 2b, using data tables found in Appendix 2A of your textbook. Which reaction produces more energy?

4. Draw Lewis dot structures for fuel 1 and fuel 2. Using your Lewis dot structures and your knowledge of chemical bonding, explain the answer that you obtained in Question 3.

CONNECTION 2
Finding Energy for the Future

Name _____

Group _____

The majority of our energy is currently supplied by fossil fuels. Because fossil fuels are nonrenewable, they will eventually be used up. As a result, scientists are looking for alternative sources of energy. Some possible alternative fuels, which are discussed in Connection 2, are hydrogen, ethanol, methane, MTBE (methyl-*tert*-butyl ether), and ETBE (ethyl-*tert*-butyl ether). When the volume of a fuel is important, a fuel's usefulness is assessed as an enthalpy density—the enthalpy of combustion per liter.

1. Calculate the enthalpy density for these five fuels using the information in your textbook, and complete the table below. You may need to use another source to find the densities and enthalpies of the fuels. (Try the World Wide Web or the *CRC Handbook*.)

Fuel	Density	Enthalpy density (kJ/L)
Hydrogen		
Ethanol		
Methane		
MTBE		
ETBE		

2. Each of these fuels has advantages and disadvantages. List some of them below.

Fuel	Advantages	Disadvantages
Hydrogen		
Ethanol		
Methane		
MTBE		
ETBE		

STUDENT COMPANION FOR CHEMISTRY

3. Based on the information that you collected in these two tables, which fuel/fuels do you think would be the best replacement for gasoline? Explain your answer.

   The amount of energy that can be produced by a fuel is related to its structure. Good fuels have relatively weak bonds. They contain elements that form strong bonds with oxygen. In general, hydrocarbons produce more energy per gram when they burn than other organic molecules. Fuels that contain oxygen, like ethanol, are useful because they burn cleaner than hydrocarbon fuels.

4. Which type of bond is weaker, a C—C, single, double, or triple bond? Compare the average bond enthalpies.

5. Octane is found in gasoline and methane, which are examples of hydrocarbon fuels. Draw Lewis structures for these fuels.

6. Ethanol, MTBE, and ETBE are examples of fuels that contain oxygen. Draw Lewis structures for these compounds. (*Hint:* See Chapter 11, Toolboxes 11.1 and 11.3 for help drawing these structures from their names. The tert-butyl group is $(CH_3)_3$—C—).

Chapter 10

Liquids and Solids

Activity 10.1 Intermolecular Forces—Guided Reading, p. 157
Activity 10.2 Crystals—CD-ROM Viewing Guide, p. 161
Activity 10.3 Unit Cell Modeling, p. 163
Activity 10.4 Phase Diagrams, p. 167
Activity 10.5 Group Application Worksheet, p. 169

ACTIVITY 10.1
Intermolecular Forces
Guided Reading

Name _____

Group _____

Read textbook sections 10.1, 10.2, and 10.3. Answer the following questions individually.

1. Some covalent bonds are polar (O—H) and some are nonpolar (I—I). Distinguish between the terms **polar** and **nonpolar.**

2. London forces hold all types of molecules together.

   a. Describe and explain the relationship that exists between molar mass and strength of London forces.

   b. Describe and explain the relationship that exists between molecular shape and strength of London forces.

3. What factors influence the strength of a dipole-dipole interaction?

4. Use an example to demonstrate how dipole-dipole interactions can cancel each other within a molecule.

5. What effect does the strength of intermolecular forces have on the boiling point of a compound? Why?

6. Describe the characteristics a molecule must possess to allow a hydrogen bond to form.

7. Is a hydrogen bond the same as a chemical bond? Why or why not?

8. Complete the questions assigned by your instructor

Answer these questions with your group.

1. Decide as a group how you would determine which of a set of bonds exhibits the highest degree of polarity. Explain your strategy.

2. Use your strategy to rank the bonds in the following compounds in terms of increasing polarity.

   _____ HCl   _____ H₂O   _____ MgS

   _____ HF   _____ CaCl₂

3. Refer to the following chart of information to answer the next questions.

Substance	Melting point, K	Boiling point, K
water	273	373
helium	3.3	4.1
carbon dioxide	—	sublimation at 194.5
silicon dioxide	2000	2500
methane	89	111

   a. Which of these substances demonstrate London forces?

   b. Does hydrogen bonding occur in any of these? If so, which ones?

c. Which substance has the weakest intermolecular forces? Why?

d. Explain how differences in bonding are related to differences in melting-point and boiling-point temperatures of silicon dioxide and carbon dioxide.

4. How can a dipole be induced in a nonpolar molecule? Diagram how this would happen on a molecular level.

ACTIVITY 10.2
Crystals
CD-ROM Viewing Guide

Name _____

Group _____

View the video demonstration on the Problem-Solving Skills CD-ROM titled "Physical Structure of Crystals," and answer the following questions.

1. What are the two elements that make up amethyst crystals?

2. Describe the difference between the structure of glass and the structure of amethyst crystals.

On the Visualization CD-ROM, under "Ions in Solution," view the construction and animation (under "Dissolving") of the cesium chloride unit cell. Build two unit cells, one with cesium as the central ion and one with chlorine as the central ion. Use your observations to answer the following questions.

3. Draw a picture of each of the unit cells in the space below.

4. What is the name for this kind of crystal structure?

STUDENT COMPANION FOR CHEMISTRY

5. When a crystal of sodium chloride forms, a characteristic shape results. Describe the shape below and include a diagram of a pattern of ions that might exist. (Note that the arrangement of ions cannot usually be predicted from the shape of a crystal.)

ACTIVITY 10.3
Unit Cell Modeling

Name _____

Group _____

In a crystal structure, the particles show a repetition of an orderly arrangement to make a structure called a **crystal lattice.** A **unit cell** is the most convenient small part of the lattice that, if repeated in three dimensions, builds the entire lattice. Depending on the lengths of the sides of the unit cell and the angles that the sides make with one another, the unit cell can be classified as one of several crystal systems. You will be building a few unit cells using the materials given to you by your instructor You will also need to select a role for each group member.

Organizer/leader _____

Model builder _____

Sketcher _____

Reporter _____

The following diagrams are found in your text.

Hexagonal close-packed

Body-centered cubic

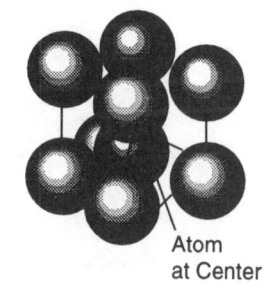

Atom at Center

Face-centered cubic

Rock salt

Cesium chloride

STUDENT COMPANION FOR CHEMISTRY

163

1. Describe the features of each type of unit cell that your team can use to distinguish among the following lattice structures.

   body-centered cubic

   face-centered cubic

   hexagonal close-packed

   rock salt

   cesium chloride

164   CHAPTER 10

2. Build the following unit cells, using the model kits or materials provided. Sketch and label a diagram of the unit cell you have built.

   a. Type: face-centered cubic

      Representative example: nickel, gold, or copper

      Coordination number: _____

   b. Type: hexagonal close-packed

      Representative example: silver chloride

      Coordination number of silver: _____

   c. Type: body-centered cubic

      Representative example: cesium iodide

      Coordination number of cation: _____

d. Type: rock salt

   Representative example: _____

3. Why are the crystal structures of copper (I) iodide (CuI) and lithium sulfide (Li₂S) different even though the anions have face-centered cubic arrangements and the cations occupy the tetrahedral holes?

ACTIVITY 10.4
Phase Diagrams

Name _____

Group _____

1. Below is a general phase diagram for a pure substance. Label each area identified.

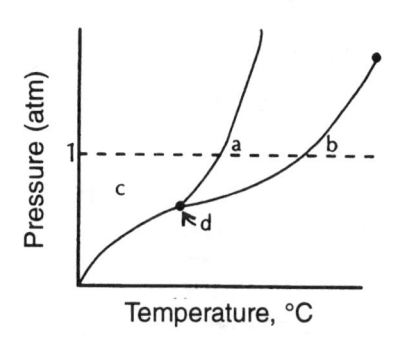

2. Explain from a molecular standpoint the phases present at the following points. Identify any phase transitions that could be taking place.

    a.

    b.

    c.

    d.

STUDENT COMPANION FOR CHEMISTRY

3. Assume that you collect the following data on a pure substance in the laboratory. Use these data and the curve-fitter program on the Problem-Solving Skills CD-ROM to produce a pressure versus temperature graph. Be sure to plot pressure on the vertical axis and temperature on the horizontal axis. Sketch or attach a print out of the resulting curve that would be obtained from the provided data.

Start Temp., °C	End Temp.,°C	Pressure, Torr	Observations in lab
0	60	460	• 0°: no evidence of vapor • 10°: solid becomes a liquid • 60°: boiling occurs • After vaporization, temp. continues to rise
0	4	228	• 4°: solid becomes vapor, with no evidence of liquid present
0	10	380	• 10°: solid, liquid, and vapor present at same time
40	85	1220	• Very high pressure; as temp. increases, a single uniform substance fills the container at 85° it is not liquid or vapor but a "fluid" intermediate

4. On your graph, label the normal freezing point, normal boiling point, and triple point.

5. On the basis of your graph, predict the state of matter that would exist under the following conditions.

   a. 190 Torr and 8°C _____ because _____

   b. 600 Torr and 50°C _____ because _____

   c. 600 Torr and 8°C _____ because _____

CHAPTER 10

ACTIVITY 10.5

Group Application Worksheet

Name _____

Group _____

Interpret the following events from daily life in terms of chemical principles you have learned.

1. Freeze-dried food is obtained by drying the food after it has been frozen. To do this, a "vacuum chamber" is typically used to apply and maintain a low air presssure to the system. Explain why the low pressure is necessary.

2. Which type of burn is more severe, a burn from boiling water or a burn from steam? Explain.

3. Liquid crystals can be used in thermometers. Propose a reason that explains why a specific temperature can be "read."

STUDENT COMPANION FOR CHEMISTRY

4. Explain why silicon and germanium make good semiconductors.

5. How can paper toweling absorb water throughout the entire towel when only one corner of it is dipped into the water?

6. Explain what is meant by "freezer burn." How is it caused?

7. Some types of glassware are often improperly called "crystal." Why would this term be incorrect? (See textbook Figure 10.11.)

Chapter 11

Organic Compounds

Activity 11.1 Organic Molecules—Guided Reading, p. 173

Activity 11.2 Shapes of Organic Molecules and Hybrid Orbitals—Group Hands-On Activity, p. 177

Activity 11.3 Functional Groups—Group Problem, p. 179

Activity 11.4 Polymers—Individual and Group Worksheet, p. 181

Activity 11.5 Chapter Summary Problem—Worksheet, p. 183

Activity 11.6 Challenge Problem, p. 187

ACTIVITY 11.1
Organic Molecules
Guided Reading

Name _____

Group _____

Hydrocarbons

Read textbook Sections 11.1–11.4 and answer the following questions.

1. Which hydrocarbons have only single bonds?

2. Which aliphatic hydrocarbons have double bonds?

3. Which aliphatic hydrocarbons have triple bonds?

4. Describe the difference between a saturated and an unsaturated hydrocarbon.

5. a. Define **aromatic hydrocarbon.**

b. Draw the structure of benzene. Indicate the hybridization of and bond angles about one of the carbon atoms. Will these be the same or different for the other carbon atoms? Look at the structural formula and model of benzene on the Problem-Solving Skills CD-ROM in the "Chem 4 Molecules Database." Compare this structure with yours.

6. List various sources of hydrocarbons.

Functional Groups

Read textbook Sections 11.5–11.10 and answer the following questions.

7. Fill in the following table. (The first entry is filled in as an example. R is used to represent any length hydrocarbon chain. Use Toolbox 11.3 to help you complete the table. Examples of organic molecules can be found on the Problem-Solving Skills CD-ROM in the "Chem 4 Molecules Database.")

Functional group name	Generic formula	Example: Condensed structural formula	Example: Name and common use
Alcohols	R—OH	CH_3OH	methanol, used as a solvent
Ethers			
Aldehydes			
Ketones			
Carboxylic acids			
Esters			
Amines			
Amides			

Return to class and answer the following questions.

8. Draw the structural formula of $CH_3CH(CH_3)CH(CH_3)CH_2CH_3$.

Name this compound, using Toolbox 11.1 as a guide.

STUDENT COMPANION FOR CHEMISTRY

9. Draw the stick formula of this compound

$$CH_3-CH_2-CH-CH-\underset{\underset{CH_2CH_3}{|}}{\overset{\overset{CH_3}{|}}{C}H_2}-CH_3$$

Name this compound, using Toolbox 11.1 as a guide.

10. For each of the functional groups listed in Question 7, indicate each type of intermolecular force present in the molecule.

ACTIVITY 11.2
Shapes of Organic Molecules and Hybrid Orbitals
Group Hands-On Activity

Name _____

Group _____

1. Using a model kit, build methane, ethane, and propane. Draw pictures of these molecules, labeling the bond angles associated with each carbon atom in your model.

   a. methane

   b. ethane

   c. propane

   d. Are all the carbon atoms in propane in a straight line? Explain your answer, using bond angles and hybridization.

STUDENT COMPANION FOR CHEMISTRY

2. a. Build ethene and draw a picture, labeling all bond angles.

   b. Are all the atoms in ethene in the same plane? Explain.

3. Build a model of ethyne, draw a picture, label bond angles, and describe its shape.

4. Can you think of any other molecules that have the same hybridization as each of the following molecules?

   a. methane

   b. ethene

   c. ethyne

5. Find an example in your textbook or on your Problem-Solving Skills CD-ROM in the "Chem 4 Molecules Database" of a molecule that has at least two of the following types of hybridization, sp, sp^2, and sp^3 in the same molecule.

ACTIVITY 11.3
Functional Groups
Group Problem

Name _____

Group _____

1. Complete the following table with your group. Use an ethyl group as your original starting point. From there, draw a structural formula of

Class of compound	Example structural formula	Rationale (how do you know that your molecule contains this functional group.)
a. an ether		
b. an alcohol		
c. a carboxylic acid		
d. an ester		
e. an amine		
f. an amide		

2. A series of models will be passed around the classroom. Refer to the label on the model and identify the functional group each contains.

I _____

II _____

III _____

IV _____

STUDENT COMPANION FOR CHEMISTRY

ACTIVITY 11.4

Polymers

Individual and Group Worksheet

Name _____

Group _____

1. Homework: Make a list of some of the polymers that you find in your home or dorm room and bring your list to class. You can identify a polymer from its recycling code or from the tag in a garment.

Source or use of polymer	Name of polymer

2. Return to class and discuss the following questions with your group members.

    a. Compare your list with those of your group members. Write formulas for the monomers of each of the polymers that you found. Identify each polymer as an addition or a condensation polymer.

Polymer name	Monomer	Type of polymer

STUDENT COMPANION FOR CHEMISTRY

b. Pick one example of an addition polymer and one example of a condensation polymer from your list and write a chemical equation showing the formation of the polymer from its monomers.

addition example

condensation example

3. Watch the demonstration "Formation of a Polymer (Nylon 6,6)" on the Problem-Solving Skills CD-ROM.

   a. Is nylon an addition or a condensation polymer?

   b. What are some uses of nylon?

   c. Write a chemical equation for the formation of nylon from its monomers.

ACTIVITY 11.5
Chapter Summary Problem
Worksheet

Name _____

Group _____

1. a. Draw the structural formula for propene.

   b. Can propene experience cis-trans isomerization? If so, draw the two structures. If not, explain why not.

2. Draw all of the molecules that have the following molecular formula: C_5H_{12}. Are these molecules structural or geometrical isomers? (Use an additional sheet of paper, if necessary.)

STUDENT COMPANION FOR CHEMISTRY

3. Draw the structures of
   a. ethanol.

   b. 2,2-dimethyl-3-ethylhexane.

   c. What kinds of intermolecular forces are present in each case?

4. Consider the following molecule.

$$CH_3 \underset{1}{\longrightarrow} \underset{2}{\overset{\overset{\displaystyle O}{\|}}{C}} \longrightarrow NH \underset{3}{\longrightarrow} CH_2 \underset{4}{\longrightarrow} CH_2 \longrightarrow OH$$

   a. Identify all the functional groups.

b. How many sigma and pi bonds are present?

c. What is the hybridization of each carbon atom?

   of the nitrogen atom?

d. What are the bond angles on carbon-2?

on carbon-4?

on nitrogen?

e. Does this molecule exhibit hydrogen bonding in its pure state?

5. The plot below shows how boiling point rises as the length of the carbon chain increases for hydrocarbons, ethers, and alcohols. Explain the differences in the boiling points of hydrocarbons, ethers, and alcohols of the *same molar mass*.

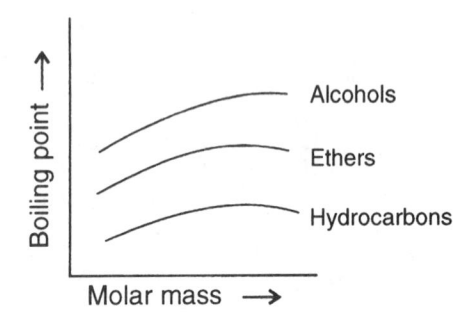

ACTIVITY 11.6

Challenge Problem

Name _____

Group _____

Consider an unknown organic compound that is found by chemical analysis to have the molecular formula, C_3H_8O.

1. On the basis of the following information, indicate a reasonable structure for this compound.

   Clue 1: It is slightly soluble in water.

   Clue 2: Upon oxidation using a mild oxidizing agent, the unknown compound converts to a compound with the formula, C_3H_6O. C_3H_6O gives a positive Tollens test.

   Draw the proposed structure and name the compound. Briefly explain your reasoning!

2. a. Write a balanced chemical equation for the complete combustion of *one* mole of the unknown organic compound.

   b. Using the chemical equation in Question 2a, the thermodynamic data in the appendices of your textbook, and relevant data from that given below, calculate the enthalpy of combustion for one mole of unknown compound.

Compound	Enthalpy of combustion at 298 K, kJ/mol
1-propanol(l)	−256.06
2-propanol(l)	−272.4
propanone(l)	−258.1
propanal(l)	−190.4

187

Chapter 12

The Properties of Solutions

Activity 12.1   Solution Basics—Worksheet, p. 191

Activity 12.2   Colligative Properties: Measuring Concentration—Guided Reading, p. 195

Activity 12.3   Colligative Properties: Vapor Pressure, Boiling-Point Elevation, Freezing-Point Depression, and Osmosis—Guided Reading, p. 199

Activity 12.4   IV Solutions—Group Problem, p. 203

Activity 12.5   CD-ROM Demonstration—Viewing Guide, p. 205

Activity 12.6   Group Challenge Problem, p. 207

ACTIVITY 12.1
Solution Basics
Worksheet

Name _____

Group _____

1. Explain how you decide which substance is a **solute** and which substance is a **solvent** in a solution.

2. Complete the following chart, filling in an example of a solution that would form when a solute, in the given state of matter, dissolves in a solvent of the given state of matter. (Assume that the *solute* is the substance written *horizontally* and the *solvent* is written *vertically* in the chart.)

SOLVENT

		Solid	*Liquid*	*Gas*
SOLUTE	Solid			Few examples are known.
	Liquid			
	Gas			

3. Water is considered to be the "universal" solvent. Explain.

STUDENT COMPANION FOR CHEMISTRY

4. Refer to the following diagram of water molecules.

   Which bonds on the diagram represent covalent bonding? _____

   Which bond on the diagram represents hydrogen bonding? _____

   ```
              C        E
   H       H - - - -O ──H
    \      /       /
   A \   / B    / D
      O         H
              H
   ```

5. Assume that water runoff from snow-packed mountains flows through a rocky stream bed from point A to point B. If the water is considered to be "pure" at point A, which ions might you expect to find dissolved in the water at point B; are they single-charged or multiple-charged ions? Explain your answer.

6. Distinguish between the terms **hydrophilic** and **hydrophobic.** Include a diagram that demonstrates your understanding of these terms.

7. Explain how shampoo cleans your hair in terms of interactions among molecules.

8. Discuss what is meant by **lattice enthalpy.** Would you expect this value to be positive or negative? Why?

9. Discuss what is meant by **enthalpy of hydration.** Would you expect this value to be positive or negative? Why?

10. Propose a relationship between the lattice enthalpy and enthalpy of hydration of a substance with an exothermic enthalpy of solution.

11. *Discuss* two possible reasons why a substance might not be soluble in water Consider all the topics discussed in this worksheet.

ACTIVITY 12.2
Colligative Properties: Measuring Concentration
Guided Reading

Name _____

Group _____

Read the material from your text about colligative properties in Section 12.11. Then answer the following questions independently.

1. **Colligative properties** are solution properties that depend only on the _____, not the _____ of the solute particles.

2. Four colligative properties of solutions are

    a.

    b.

    c.

    d.

3. Distinguish between the terms **molarity** and **molality**. Discuss a situation in which each concentration unit would be useful.

4. What information do you need to calculate the mole fraction of a solute in a solution?

STUDENT COMPANION FOR CHEMISTRY   195

5. Calculate the mole fraction of propanol (C₃H₇OH) in a solution of 40.0 g of propanol and 40.0 g of water.

6. How do you convert from mole fractions to molality?

7. How do you convert from molality to mole fractions?

8. Complete the homework problems assigned by your instructor.

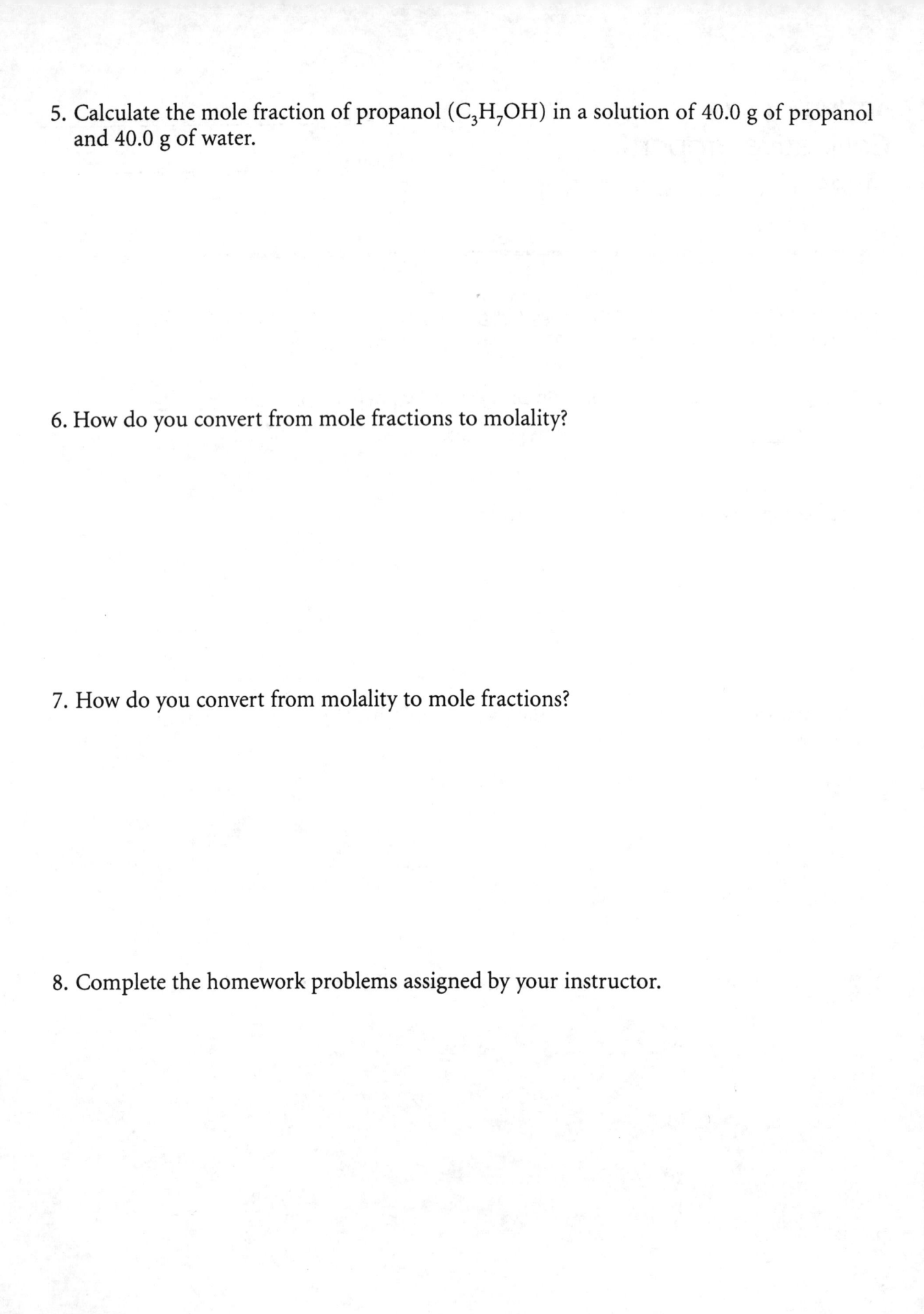

Complete the following question within your group.

9. In lab, a solution is prepared by mixing 2.00 g of ethanol (C_2H_5OH) with 125.0 mL of distilled water. The density of pure water at this temperature is 1.0 g/mol. The final volume of the resulting solution is 126.0 mL. What information do you need in order to determine

   a. the molarity of this solution?

   b. the mass percentage of each component?

   c. the mole fraction of each component?

   d. the molality of this solution?

   Use the information given to calculate each of the components above.

STUDENT COMPANION FOR CHEMISTRY

ACTIVITY 12.3

Name _____

Colligative Properties: Vapor Pressure, Boiling-Point Elevation, Freezing-Point Depression, and Osmosis

Group _____

Guided Reading

Read Sections 12.12, 12.13, and 12.14 in your text. Then answer the following questions independently.

1. Explain **Raoult's law,** using your own words. Include the equation as part of your explanation.

2. Why must the solutes that lower the vapor pressure and raise the boiling point of a solvent be **nonvolatile**?

3. Refer to the strategy from Example 12.9 "Using Raoult's law." What information is needed to solve problems such as this?

4. Ideal solutions are those that follow Raoult's law at all concentrations. What are nonideal solutions and when do they follow Raoult's law?

5. Explain *why*, when a nonvolatile solute is added to a pure solvent,

   a. the vapor pressure is lowered.

   b. the boiling point is raised.

   c. the freezing point is lowered.

6. Explain why the van't Hoff *i* factor is helpful only when considering dilute solutions.

7. What is **osmosis**?

8. a. Write the equation that can be used to determine osmotic pressure. Label all components with their units.

   b. What is **reverse osmosis** and when is it used?

9. Complete the questions assigned by your instructor

Complete the following question within your group.

10. Assume that you have 2.6 g of an unknown solute dissolved into 250.0 g of cyclohexane solvent. How would you determine the molar mass of the unknown solute from the freezing point of cyclohexane?

    As a group, write the strategy or sequence of steps needed in order to solve this problem.

    Step 1:

STUDENT COMPANION FOR CHEMISTRY

Step 2:

Step 3:

Assign each person in your group one of the following situations to work on, using the strategy you developed.

What would the molar mass of this unknown solute be if the freezing point of the solution were determined to be

a. 6.4°C _____

b. 6.3°C _____

c. 6.2°C _____

d. 6.1°C _____

(Circle the letter of the portion you were responsible for calculating.)

In general, what relationship is there between molar mass of the solute and the amount of depression of the freezing point?

ACTIVITY 12.4
IV Solutions
Group Problem

Name _____

Group _____

Each group member will be given a card that contains a piece of information needed to solve a problem. Once you receive your card, read the information provided. Write the step-by-step strategy that you use to come up with your answer.

1. Explain the effect that a 0.32 m solution of NaCl would have on the body if it were given to you as an intravenous (IV) solution.

ACTIVITY 12.5
CD-ROM Demonstration
Viewing Guide

Name _____

Group _____

View the "Molecular Level Animations" on the Visualization CD-ROM about NaCl solution and dissolution of NaCl. Answer the following questions after viewing the sequences.

1. Explain, using a series of diagrams, what you might see if you were to look at a beaker of water while increasing the temperature.

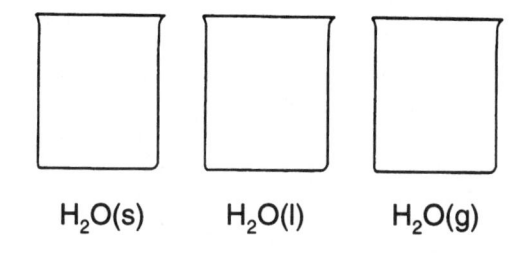

   $H_2O(s)$ $H_2O(l)$ $H_2O(g)$

2. Explain how increasing temperature affects distance between water molecules as well as their motion.

3. What happens on the molecular level when water boils and evaporates?

STUDENT COMPANION FOR CHEMISTRY

4. Diagram what happens when sodium chloride is added to water. Label all components as well as the solute and solvent.

5. Explain (in words) what is happening in Question 4.

ACTIVITY 12.6
Group Challenge Problem

Name _____

Group _____

A new product called Drain Away is being developed. Like many drain openers, this crystalline solid contains sodium hydroxide as its main ingredient. Sodium hydroxide has a negative enthalpy of reaction with grease.

1. Would you expect the temperature to rise or fall when NaOH is added to a greasy drain? Explain.

In addition, Drain Away also contains a small amount of aluminum which also reacts with the sodium hydroxide to form $NaAlO_2$ and hydrogen gas in an exothermic reaction.

2. Write a balanced equation for this reaction.

Assume that a 425.0 g container of Drain Away has 390.0 g of sodium hydroxide and 35.0 g of aluminum shavings.

3. What reactant limits the reaction?

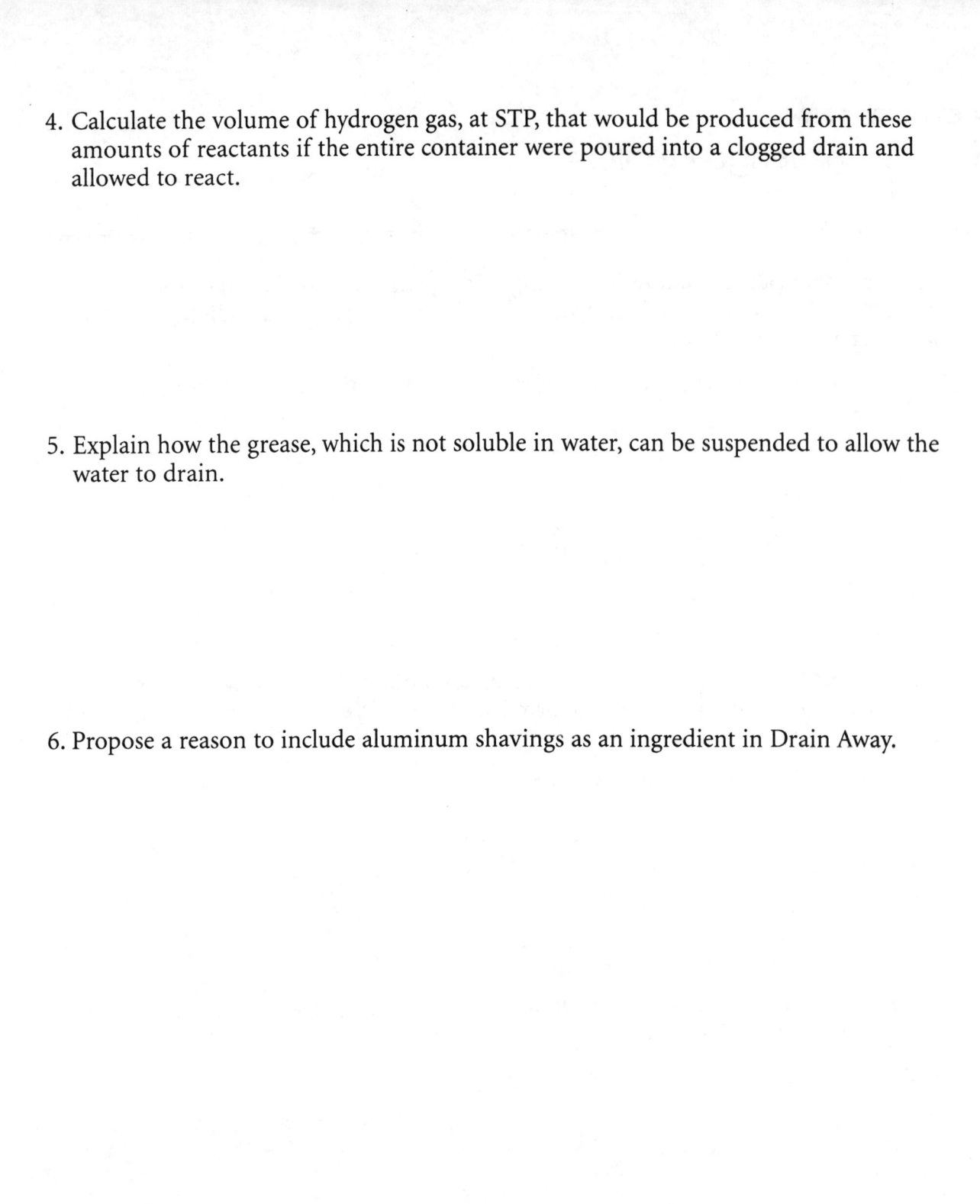

4. Calculate the volume of hydrogen gas, at STP, that would be produced from these amounts of reactants if the entire container were poured into a clogged drain and allowed to react.

5. Explain how the grease, which is not soluble in water, can be suspended to allow the water to drain.

6. Propose a reason to include aluminum shavings as an ingredient in Drain Away.

CONNECTION 3
Sports Drinks

Names A:_____   C:_____

B:_____   D:_____

Group _____

Each person in your group (identified A, B, C, and D) will analyze the label information found on a popular sports drink. Each person will be responsible for his or her own calculations and answers and as a group will decide which drink would be the best choice based upon these calculations.

Situation: Your community has decided to sponsor an athletic competition to raise money to support local charities. You have decided to compete in this local race which consists of a 10 km run, followed by a 1 mile swim, and a 50 km bike race. While you have trained appropriately, you want to make sure that you get the needed electrolytes for during the race not only to prevent dehydration, but also to perform at your best possible level.

Background: Your body stores carbohydrates as a polymer called glycogen in your muscles as well as your liver. Glycogen is broken down into simple glucose units and released into the bloodstream, as it is needed for energy. High intensity exercise can deplete the glycogen reserves within 60–90 minutes. Blood sugar levels drop and lactic acid builds up. Fatigue and pain are often due to increased lactic acid in the muscles. While working, muscles also generate a large amount of heat that is dissipated by water in your system. Sweating and evaporation are important, but sweating away more than 2% of your body weight can be dangerous to your health in part because it contains important cations, such as potassium and sodium. The ability of a solute, such as potassium or sodium, to enter the cell is determined by the concentration of solute inside versus outside of the cell. Osmosis involves the diffusion of water through the membrane from a solution of higher solute concentration to one with a lower solute concentration.

As a group, respond to the following before you complete your calculations:

1. Would you expect lactic acid to be considered a weak or a strong acid? Explain your reasoning.

2. Write the balanced chemical equation to describe the combustion reaction that occurs when glucose, $C_6H_{12}O_6$, reacts with oxygen gas.

3. Would this reaction have a positive or negative enthalpy change? Explain how you decided your answer.

4. What is an electrolyte?

5. Why would you want an electrolyte to drink after physical activity?

6. Distinguish between **iso**tonic, **hypo**tonic, and **hyper**tonic solutions and describe their effect on living cells. These terms are important in animal cell biology because osmotic imbalances cause the cell membranes to shrink or swell depending on solute concentrations. Plant cells contain cell walls that make the cells more rigid.

The following information was taken from the labels of four different sports drinks found in the grocery store.

	Sample A	Sample B	Sample C	Sample D
Bottle size	946 mL (32 oz)	473 mL (16 oz)	946 mL (32 oz)	325 mL (11 oz)
Servings per bottle	4	1	2	1
Calories per serving (kcal)	80	400	250	220
Sodium per serving	80.0 mg	55.0 mg	55.0 mg	220.0 mg
Potassium per serving	50.0 mg	99.0 mg	50.0 mg	530.0 mg
Total carbohydrates per serving*	20.0 g	100.0 g	20.0 g	38.0 g
Glucose per serving*	0	30.0 g	19.0 g	33.0 g
Drink density	1.1 g/mL	1.3 g/mL	1.2 g/mL	1.4 g/mL

*Total carbohydrates per serving include grams of sugar per serving.

Each person will determine the following about their assigned sports drink:

- The mass of sodium and potassium per 250.0 mL of drink

- Osmolality is a term used to describe the number of solute particles dissolved in 1 liter of water and is reflected in the solution's ability to cause osmosis. One *osmol* = 1 mole of **non-penetrating** solute (such as these cations) per liter of water. A *milliosmol* is a common concentration unit when discussing body fluids. Calculate the number of milliosmols (mosm) due to sodium and potassium cations alone in your drink assuming that the solvent in the sports drink is only water.

- What would happen if this concentration of potassium and sodium that you calculated were put into a human body, which typically has an intercellular sodium concentration of 142.0 mosm and a potassium concentration of 4.5 most? (*Hint:* Consider if it is hypotonic, hypertonic, or isotonic.)

As a group, compare your calculations and identify which drink would be the best for your situation. Explain your reasoning.

Chapter 13

The Rates of Reactions

Activity 13.1   Introduction to Reaction Rates—Guided Reading, p. 213
Activity 13.2   Integrated Rate Laws—Group Problem, p. 217
Activity 13.3   Collision Theory—Lecture Demonstration and Group Discussion, p. 219
Activity 13.4   An Iodine Clock Reaction—CD-ROM Demonstration and Viewing Guide, p. 223
Activity 13.5   Group Challenge Problem, p. 225

ACTIVITY 13.1
Introduction to Reaction Rates
Guided Reading

Name _____

Group _____

Read Sections 13.1 and 13.2 in your textbook.

1. Define the following terms.

   a. rate of reaction

   b. tangent line

   c. instantaneous rate

2. Work the textbook exercises assigned by your instructor.

Read Section 13.3.

3. Write a general expression for a rate law.

What is *k*?

4. Give one example of each of the following reactions.
   a. a reaction that is zero order overall

   b. a reaction that is first order overall

   c. a reaction that is second order overall

Read Section 13.4.

5. Do reaction orders always have to be positive integers? Explain.

6. Develop a strategy for determining rate law expressions, using initial rate data.

7. Using your strategy, work textbook Exercise 13.21 in your groups.

ACTIVITY 13.2
Integrated Rate Laws
Group Problem

Name _____

Group _____

1. Answer the following questions after examining the data presented in the accompanying plots (at 1000 K) for the reaction

$$2 N_2O(g) \rightarrow 2 N_2(g) + O_2(g)$$

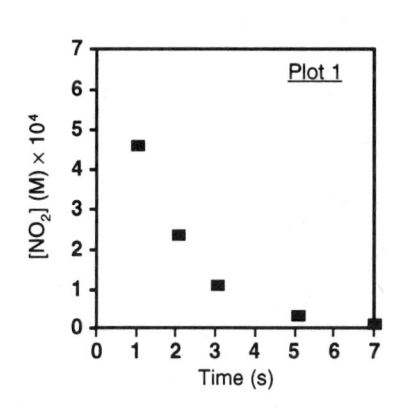

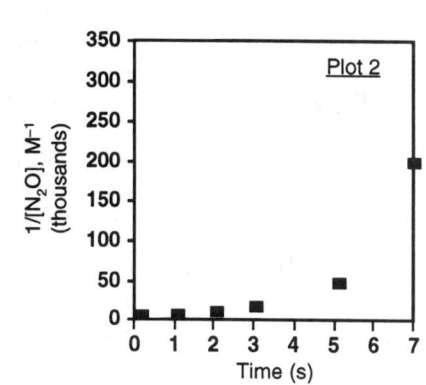

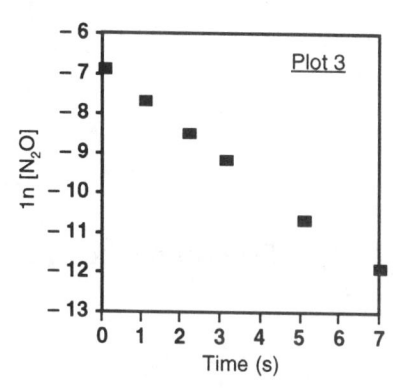

Plot 1: $[N_2O]$ (M) × 10^4 vs. Time (s)

Plot 2: $1/[N_2O]$, M^{-1} (thousands) vs. Time (s)

Plot 3: $\ln [N_2O]$ vs. Time (s)

a. What is the overall order of the reaction? Explain your answer.

b. Determine the value of the rate constant at 1000 K.

c. What is the half-life for this reaction?

2. Use the curve-fitter program on your CD-ROM and the following data at 600 K for the reaction, 2 HI(g) → H$_2$(g) + I$_2$(g), to answer the following questions.

[HI], M	Time, min
0.50	0
0.488	100
0.457	300
0.417	700
0.373	1200
0.315	2000

a. Determine the overall order for this reaction.

b. What is the value of the rate constant for this reaction at 600 K?

c. Calculate the half-life of this reaction at 600 K.

ACTIVITY 13.3
Collision Theory
Lecture Demonstration and Group Discussion

Name _____

Group _____

1. Watch the lecture demonstration of the reaction of magnesium metal with hydrochloric acid and record your observations below.

    a. What happens to the rate of reaction as the concentration of hydrochloric acid is increased?

    b. What happens to the rate of reaction as the temperature is increased?

    c. What happens to the rate of reaction as the temperature is decreased?

    d. Compare the rates of reaction for solid magnesium strips and powdered magnesium.

STUDENT COMPANION FOR CHEMISTRY

2. Write a balanced chemical equation for the reaction shown in the demonstration.

3. Use collision theory to explain your observations recorded in Question 1.

   a.

   b.

   c.

   d.

4. Use collision theory to explain the effect of a catalyst on a chemical reaction.

5. Some catalysts increase the rate of a chemical reaction by orienting the reactant particles in the most favorable position for reaction to occur. Do you think that the reaction rate would increase or decrease if most of the magnesium atoms hit the chlorine end of the HCl? if they hit the hydrogen end of the HCl?

ACTIVITY 13.4
An Iodine Clock Reaction
CD-ROM Demonstration and Viewing Guide

Name _____

Group _____

Watch the iodine clock reaction demonstration on your Problem-Solving Skills CD-ROM and answer the following questions.

a. What happens to the rate of reaction as the concentrations of reactants vary? Make a table and record your observations below.

b. Do your observations agree with the predictions of collision theory?

c. Design an experiment to study the effect of temperature on the rate of the iodine clock reaction. Predict the qualitative results of your experiments and explain your answer.

ACTIVITY 13.5
Group Challenge Problem

Name _____

Group _____

Many organic bromides are hydrolyzed by hydroxide ions in ethanol to form alcohols. But how does this happen? Does the hydroxide ion attack the molecule from behind, displacing the bromide in one step?

$$OH^- + R-X \rightarrow HO-R + X^-$$

Or does the molecule ionize first and then bond to the hydroxide in this two-step process?

$$R-X \rightarrow R^+ + X^-$$

$$R^+ + OH^- \rightarrow R-OH$$

Using the initial rate data below, collected for *n*-butyl bromide and *t*-butyl bromide,

a. find the rate law for the hydrolysis of each bromide.

b. for each reaction suggest which of the above mechanisms is consistent with the experimental data.

Reactant	[RX], M	[OH⁻], M	Initial rate, M/s
n-Butyl bromide	0.001	0.01	3.0×10^{-3}
$CH_3-CH_2-CH_2-CH_2-Br$	0.002	0.01	6.0×10^{-3}
	0.001	0.03	9.0×10^{-3}
t-Butyl bromide	0.001	0.01	1.2×10^{-2}
CH_3 | CH_3-C-CH_3 | Br	0.002	0.01	2.4×10^{-2}
	0.001	0.03	1.2×10^{-2}

STUDENT COMPANION FOR CHEMISTRY

Chapter 14

Chemical Equilibrium

Activity 14.1   Writing and Understanding Equilibrium Constant Expressions—Worksheet, p. 229

Activity 14.2   Equilibrium Tables—Guided Reading, p. 231

Activity 14.3   Le Chatelier Principle—Group Problem and Lecture Demonstration, p. 233

Activity 14.4   The Temperature Dependence of the Equilibrium Constant—CD-ROM Viewing Guide, p. 235

Activity 14.5   Calculating Equilibrium Constants—Group Problem, p. 237

ACTIVITY 14.1
Writing and Understanding Equilibrium Constant Expressions
Worksheet

Name _____

Group _____

---

1. Write the equilibrium constants for the following reactions, using the law of mass action.

   a. $2\ BrCl(g) \rightleftharpoons Br_2(g) + Cl_2(g)$

   b. $F_2(g) \rightleftharpoons 2\ F(g)$

   c. $NH_4(NH_2CO_2)(s) \rightleftharpoons 2\ NH_3(g) + CO_2(g)$

   d. $2\ HBr(g) \rightleftharpoons H_2(g) + Br_2(g)$

   e. $4\ NH_3(g) + 5\ O_2(g) \rightleftharpoons 4\ NO(g) + 6\ H_2O(g)$

   f. $HC_2H_3O_2(aq) + H_2O(l) \rightleftharpoons C_2H_3O_2^-(aq) + H_3O^+(aq)$

   g. $Ni(s) + 4\ CO(g) \rightleftharpoons Ni(CO)_4(g)$

2. Are products or reactants favored at equilibrium for the following reactions? Explain your answers.

   a. $2\ BrCl(g) \rightleftharpoons Br_2(g) + Cl_2(g)$ $K_c = 4.0 \times 10^{18}$ at 500 K

   b. $2\ HBr(g) \rightleftharpoons H_2(g) + Br_2(g)$ $K_c = 7.7 \times 10^{-11}$ at 500 K

STUDENT COMPANION FOR CHEMISTRY

c. $HC_2H_3O_2(aq) + H_2O(l) \rightleftharpoons C_2H_3O_2^-(aq) + H_3O^+(aq)$ $K_c = 1.8 \times 10^{-5}$

3. a. Calculate K_c for the reverse of reaction 2a.

b. Calculate K_p for reaction 2b.

c. If K_c is 0.28 at 500 K for $2\ HD(g) \rightleftharpoons H_2(g) + D_2(g)$, calculate K_c at 500 K for $HD(g) \rightleftharpoons 1/2\ H_2(g) + 1/2\ D_2(g)$.

# ACTIVITY 14.2
Equilibrium Tables
Guided Reading

Name _____

Group _____

Read textbook Sections 14.6 through 14.8.

1. List the steps involved in setting up an equilibrium table (see Toolbox 14.1).

2. Show how you use the quadratic formula to solve equations like the following example.
$$4.0x^2 + 5.0x - 3.0 = 0.0$$

3. How do you decide which root of the quadratic equation to use?

4. How do you decide whether you can simplify an equilibrium problem so that you do not have to solve a quadratic or a cubic equation?

STUDENT COMPANION FOR CHEMISTRY

231

5. Use your equilibrium table procedure from Question 1 to solve the following homework problems assigned by your instructor.

You may want to try using the "equilibrium calculator" on your Problem-Solving Skills CD-ROM to solve these problems.

Return to class and complete the following problem in your groups.

6. The gas BrCl decomposes according to the equilibrium

$$2\ BrCl(g) \rightleftharpoons Br_2(g) + Cl_2(g)$$

for which $K_c = 5.0$ at 1000 K. If pure BrCl at a concentration of 0.10 mol/L is allowed to decompose, what is the composition of the equilibrium mixture? *Hint:* Set up an equilibrium table.

ACTIVITY 14.3

Le Chatelier Principle

Group Problem and Lecture Demonstration

Name _____

Group _____

1. As a group, develop a reasonable explanation of the Le Chatelier principle.

2. What information/data do you need to solve equilibrium problems?

3. Watch the lecture demonstration and describe what you observe. Be specific.

4. Write an equation for the reaction you observed in Question 3.

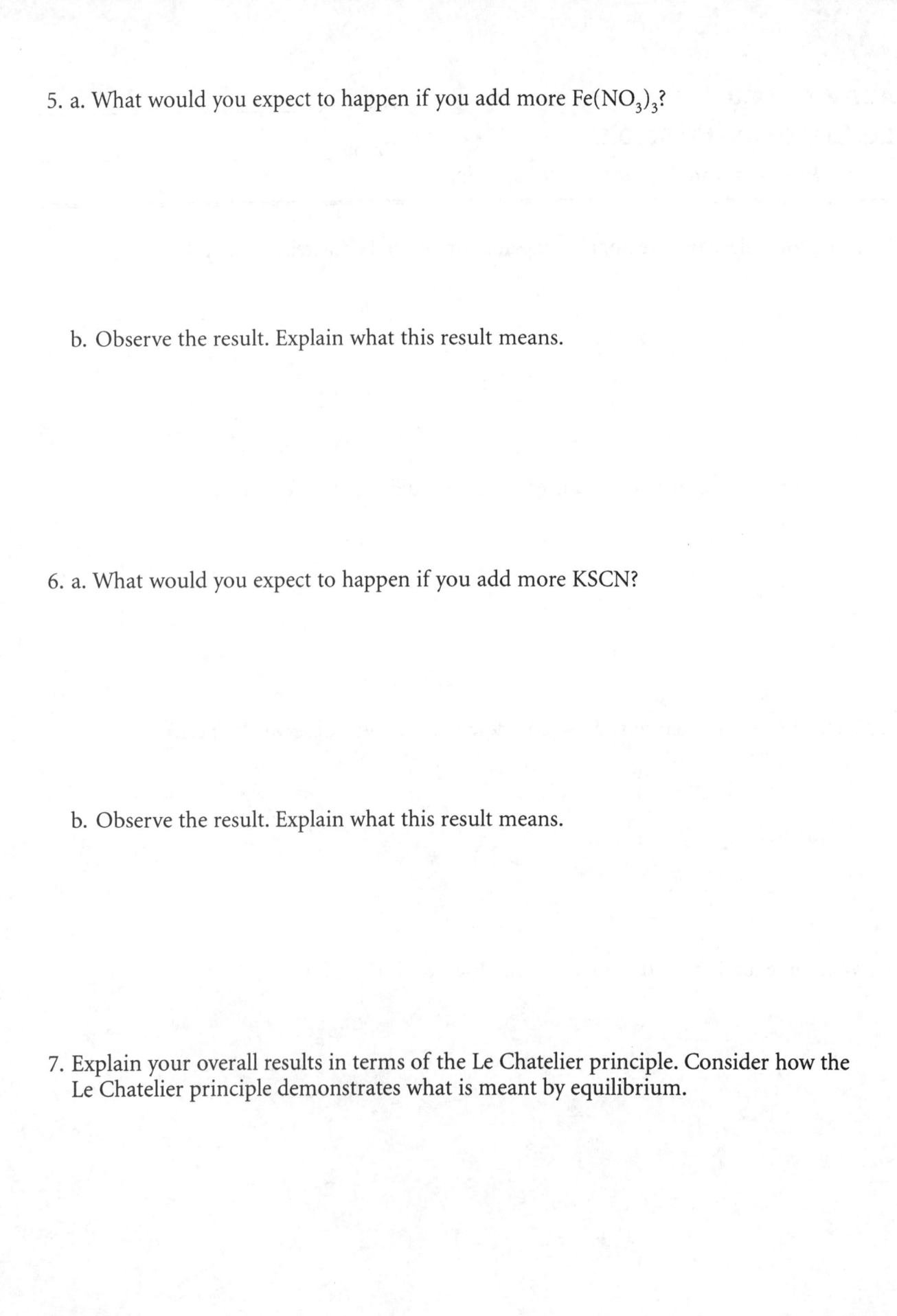

5. a. What would you expect to happen if you add more Fe(NO$_3$)$_3$?

   b. Observe the result. Explain what this result means.

6. a. What would you expect to happen if you add more KSCN?

   b. Observe the result. Explain what this result means.

7. Explain your overall results in terms of the Le Chatelier principle. Consider how the Le Chatelier principle demonstrates what is meant by equilibrium.

ACTIVITY 14.4
The Temperature Dependence of an Equilibrium Constant
CD-ROM Viewing Guide

Name _____

Group _____

Watch the Problem-Solving Skills CD-ROM demonstration "The Temperature-Dependent Equilibrium of Cobalt Complex Ions" and answer the following questions.

1. Which cobalt complex ion forms a pink solution?

2. Which cobalt complex ion forms a blue solution?

3. Write a balanced chemical equation representing the equilibrium reaction that is occurring in this demonstration.

4. How do the following changes affect the color of the solution?

   a. heating

   b. cooling

STUDENT COMPANION FOR CHEMISTRY

5. What color is the equilibrium solution at room temperature? Explain why the solution is this color.

6. Is the reaction described in Question 3 endothermic or exothermic? Explain your answer on the basis of the results of this demonstration.

ACTIVITY 14.5
Calculating Equilibrium Constants

Group Problem

Name _____

Group _____

1. In this problem, you will calculate the equilibrium constant from equilibrium concentrations for a reaction that will have the following generic form

$$2 \text{ A}(g) \rightleftharpoons 2 \text{ B}(g) + \text{C}(g)$$

    Start by discussing the strategy for calculating the equilibrium constant from equilibrium concentrations for a reaction. Record your group strategy below. *Hint:* You may want to refer to your guided reading.

2. Divide your group in two. Half of the group will work problem (a) and the other half, problem (b).

    a. 0.10 mol of H_2S gas decomposed at 1132°C in a 1.0-L container according to the following reaction

    $$2 \text{ H}_2\text{S}(g) \rightleftharpoons 2 \text{ H}_2(g) + \text{S}_2(g)$$

    Calculate K_c for this reaction if 0.00285 mol of hydrogen gas remains at equilibrium.

b. 0.010 mol NOCl decomposes at 500 K in a 1.0-L container, according to the following reaction

$$2\ NOCl(g) \rightleftharpoons 2\ NO(g) + Cl_2(g)$$

Calculate K_c for this reaction if 0.010 mol of NOCl is 9.0% dissociated at equilibrium.

3. Exchange papers and correct each other's work. After you are done, discuss what you learned.

Chapter 15

Acids and Bases

Activity 15.1   Acids and Bases—Worksheet, p. 241
Activity 15.2   pH and pOH—Guided Reading, p. 245
Activity 15.3   Ionization of Weak Acids and Bases—Group Problem, p. 249
Activity 15.4   Ammonia Fountain—CD-ROM Viewing Guide, p. 251
Activity 15.5   Stoichiometry—Group Problem, p. 255
Activity 15.6   Polyprotic Acids and Bases—Group Challenge Problem, p. 257

ACTIVITY 15.1
Acids and Bases
Worksheet

Name _____

Group _____

1. What are three chemical characteristics of acids?

    a.

    b.

    c.

2. What are three chemical characteristics of bases?

    a.

    b.

    c.

3. In the Brønsted-Lowry system, what is meant by the terms **conjugate acid** and **conjugate base**?

4. Identify the **acid, base, conjugate acid,** and **conjugate base** in each of the following equations.

    a. $HA + H_2O \rightleftharpoons H_3O^+ + A^-$ (generic formula)

STUDENT COMPANION FOR CHEMISTRY 241

b. $HNO_3(aq) + NaOH(aq) \rightleftharpoons H_2O(l) + NaNO_3(aq)$

c. $NaHCO_3(aq) + HCl(aq) \rightleftharpoons H_2CO_3(aq) + NaCl(aq)$

d. $NH_2NH_3^+(aq) + H_2O(l) \rightleftharpoons H_3O^+(aq) + NH_2NH_2(aq)$

5. Write the chemical quation in which water acts as an acid.

6. Write a chemical equator for a reaction in which water acts as a base.

7. What term is used to describe a substance that can act as either an acid or a base?

8. How do you distinguish between a strong acid and a weak acid? Cite an example of each.

9. How do you distinguish between a strong base and a weak base? Cite an example of each.

10. Explain the relationship that exists between the strength of an acid and the strength of its conjugate base.

11. Use the table to rank the following acids in terms of decreasing strength.

Formula	Acid name	K_a
HCOOH	Formic acid	1.8×10^{-4}
CH_3COOH	Acetic acid	1.8×10^{-5}
C_6H_5COOH	Benzoic acid	6.5×10^{-5}

12. Use the table to rank the following bases in order of increasing strength.

Formula	Base name	K_b
NH_3	ammonia	1.8×10^{-5}
NH_2NH_2	hydrazine	1.7×10^{-6}
NH_2OH	hydroxlamine	1.1×10^{-8}

13. How are the Lewis definitions of acids and bases similar to the Brønsted-Lowry definitions?

STUDENT COMPANION FOR CHEMISTRY

ACTIVITY 15.2
pH and pOH
Guided Reading

Name _____

Group _____

Read Sections 15.4 and 15.5 in your text. Then answer the following questions individually.

1. What is the formula used to calculate pH from the hydronium ion concentration?

2. Discuss how pH changes as the concentration of the hydronium ion increases or decreases.

3. Rainwater usually has dissolved carbon dioxide in it; therefore, its pH is approximately 5.4. A rainwater sample was taken in a national forest area and the measured pH was 3.4. Compare the hydronium ion concentrations of these samples.

4. Refer to Example 15.3. Write down the three steps required to calculate the pH of a solution.

5. Show how you can calculate pH from pOH.

6. Determine the pH and discuss the acidity or alkalinity of solutions with the following pOHs.

   a. pOH = 2.00

   b. pOH = 9.65

   c. pOH = 7.01

7. Complete the problems assigned by your instructor.

After you have completed the individual portion, answer the following questions in your group.

8. Each group member will be responsible for calculating his or her "assigned" component. In addition, group members should agree on all answers before he or she turns this component in to the instructor.

   Group member assigned to A: _____

   Group member assigned to B: _____

   Group member assigned to C: _____

   Group member assigned to D: _____

   a. Assume that these substances are strong acids and bases that ionize completely in water. Calculate the pH of these aqueous solutions.

   A. 0.0001 M HCl

   B. 0.001 M HCl

   C. 0.010 M HCl

   D. 0.100 M HCl

   b. Calculate the pOH and pH of these aqueous solutions.

   A. 0.0001 M KOH

   B. 0.001 M KOH

   C. 0.010 M KOH

   D. 0.100 M KOH

   c. How does the pH and pOH of a solution change as its acidity increases?

ACTIVITY 15.3
Ionization of Weak Acids and Bases
Group Problem

Name _____

Group _____

1. When working with weak acids and bases, you should set up an equilibrium table before you begin. Why is this important with weak acids and bases but not strong ones?

2. Once you have an equilibrium table, develop a strategy that you, as a group, can use to determine the answer to these type of problems.

3. Use your strategy to solve the following problem.

    Calculate the percentage of HCOOH molecules that are ionized in 0.10 M and 0.01 M solutions of HCOOH(aq) with $K_a = 1.77 \times 10^{-4}$. The equation for the ionization is

    $$H_2O(l) + HCOOH(aq) \rightleftharpoons H_3O^+(aq) + HCOO^-(aq)$$

STUDENT COMPANION FOR CHEMISTRY

4. Use the Le Chatelier principle to explain how your answers support expected results.

ACTIVITY 15.4
Ammonia Fountain
CD-ROM Viewing Guide

Name _____

Group _____

Watch the Problem-Solving Skills CD-ROM demonstration titled "An Ammonia Fountain" and answer the following questions.

1. Explain what is meant by an **indicator** and what indicators are used for.

2. Phenolphthalein is a colorless weak acid with a conjugate base that turns the color _____ in solutions with a pH above _____.

3. Two broad-spectrum indicators are universal indicator and pH paper. Explain how these indicators can turn a variety of colors, depending on the pH of the solution tested.

4.

Indicator	Color change	pH
Methyl orange	red to yellow	3.0–4.4
Phenol red	yellow to red	6.4–8.2
Thymol blue	red to yellow	1.2–2.8
	yellow to blue	8.0–9.6

Assume that you are designing an experiment to determine the amount of acid found in a carbonated beverage. Refer to the tabulated information and answer the following questions.

a. What do you need to consider when selecting an indicator for a solution?

b. Which indicator would you select for a solution with a pH = 2? A pH = 5? Explain.

5. Diagram three molecules of ammonia. Indicate the types of intermolecular forces present between these molecules.

252

CHAPTER 15

6. Why is phenolphthalein used as an indicator in this demonstration?

7. Discuss what you observed during the demonstration and explain why this occurs.

8. What is left over in the flask? How do you know?

ACTIVITY 15.5
Stoichiometry
Group Problem

Name _____

Group _____

Because of your success in chemistry class, a local antacid developer has hired your group to determine the effectiveness of their antacid, Burn-No-More. You are being well paid for your time and efforts so you are very careful with your data gathering. In addition to determining the comparative effectiveness, you are also going to be paid to endorse the product if your data support its success.

You know that there are two typical reactions that occur when antacids are used. A product may contain carbonate ions or hydroxide ions such as aluminum hydroxide. Assume that Burn-No-More contains calcium carbonate as its active ingredient. Write the balanced chemical equation for the reaction of calcium carbonate with HCl(aq).

Your data table.

Antacid brand	Concentration of HCl	Amount of antacid (in grams) needed to completely react with HCl
X	1.0	26.0
Y	1.0	51.0
Z	1.0	74.9

All three samples contained the same active ingredient and each sample was ground up and dissolved in enough water to make 100.0 mL of antacid solution.

1. What is the concentration of calcium carbonate in X, Y, and Z?

STUDENT COMPANION FOR CHEMISTRY

2. The suggested retail price and dosage of each are also recorded.

   Brand X $1.25 for a package of 24 tablets   dose = 2 tablets

   Brand Y $0.99 for a package of 12 tablets   dose = 2 tablets

   Brand Z $2.29 for a package of 36 tablets   dose = 1 tablet

   Calculate the price per dose for each brand.

3. Which would be a "best buy"? Explain.

4. You find out after testing that Brand X is Burn-No-More. Do you support/endorse this product? Explain.

ACTIVITY 15.6
Polyprotic Acids and Bases
Group Challenge Problem

Name _____

Group _____

Problem 1

Consider the following as an acid ionization equilibria.

a. $H_3PO_4(aq) \rightleftharpoons H^+(aq) + H_2PO_4^-(aq)$ $K_{a1} = 7.5 \times 10^{-3}$

b. $H_2PO_4^-(aq) \rightleftharpoons H^+(aq) + HPO_4^{2-}(aq)$ $K_{a2} = 6.2 \times 10^{-8}$

c. $HPO_4^{2-}(aq) \rightleftharpoons H^+(aq) + PO_4^{3-}(aq)$ $K_{a3} = 4.8 \times 10^{-13}$

1. Why is H_3PO_4 called a polyprotic acid?

2. Which substance acts **only** as an acid?

3. Which substance acts **only** as a base?

4. Which substances can act **either** as an acid or as a base?

5. What information do you need to calculate the pH of a 1.0 M solution of H_3PO_4?

6. Calculate this pH.

Problem 2

Assume that you have a flask of 0.1 M barium hydroxide and a flask of 0.1 M sulfuric acid. You decide to combine 100.0 mL of each of these together.

1. Write a balanced chemical equation for this reaction.

2. Write this equation as a net ionic equation.

3. Determine the pH of the final mixture.

4. Determine the pOH of the final mixture.

… Chapter 16

Aqueous Equilibria

Activity 16.1 Salts as Acids and Bases—Worksheet, p. 261
Activity 16.2 Titrations—Guided Reading and Group Worksheet, p. 263
Activity 16.3 Indicators as Weak Acids—CD-ROM Viewing Guide, p. 267
Activity 16.4 Buffers—Group Challenge Problem with Individual Accountability, p. 269
Activity 16.5 Solubility—Guided Reading and Group Problem, p. 271
Activity 16.6 Precipitates—Group Problem, p. 273

ACTIVITY 16.1
Name _____

Salts as Acids and Bases
Group _____
Worksheet

1. Will aqueous solutions of the following salts be acidic, basic, or neutral? (Use Tables 16.1 and 16.2 in your textbook to verify your predictions.)

   a. $FeCl_3$

   b. NH_4NO_3

   c. $NaCN$

   d. $NaClO_4$

   e. CH_3NH_3Br

2. Calculate the pH of a 0.25 mol/L solution of NH_4NO_3.

3. Predict the direction of the pH change for the following solutions:

   a. $NaC_2H_3O_2$ is added to an aqueous solution of $HC_2H_3O_2$.

   b. NH_4Cl is added to an aqueous solution of NH_3.

ACTIVITY 16.2

Name _____

Titrations

Group _____

Guided Reading and Group Worksheet

Read Sections 16.4 and 16.5 in your textbook.

1. 200.0 mL of 0.10 M HCl(aq) is prepared.

    a. List all species (molecules and ions) present in the solution and give the concentration of each, except for H_2O.

    b. What is the pH of the solution?

    c. What volume of 0.30 M NaOH (aq) would be required to neutralize the acid?

    d. What is the pH of the neutralized solution?

    e. Give the concentrations of all species in the neutralized solution.

2. Work the textbook exercises assigned by your instructor.

STUDENT COMPANION FOR CHEMISTRY

3. Suppose we add 50.0 mL of 0.20 M NaOH(aq) to the *original* HCl solution in Question 1.

   a. How many moles of NaOH have been added?

   b. How many moles of HCl react?

   c. Which reactant is limiting? Which is in excess? How many moles of the excess reactant are left after the reaction is complete?

   d. Calculate the [H$_3$O$^+$] and the pH of the solution that exists after the reaction is complete.

4. Now add the following volumes of 0.20 M NaOH(aq) to the *original* solution in Question 1, and calculate the final pH.

   a. 70.0 mL

   b. 90.0 mL

c. 110.0mL

d. 130.0 mL

5. Return to class and plot the pH of the system as a function of volume of added NaOH. Identify the stoichiometric point.

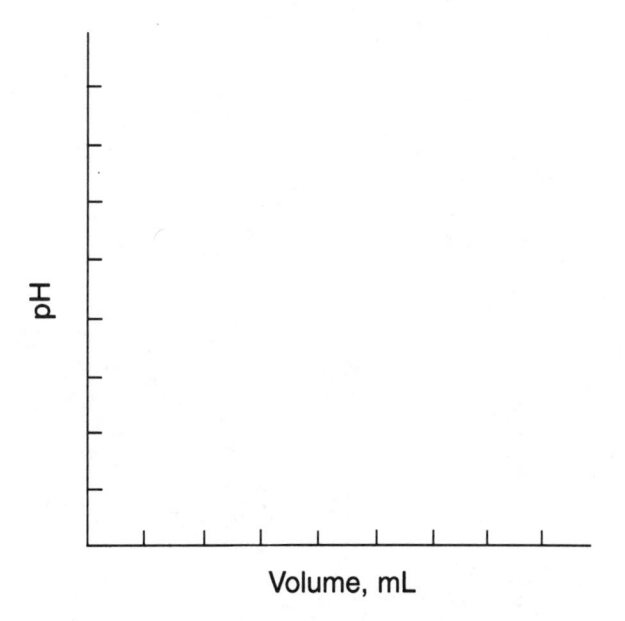

pH | Volume, mL

ACTIVITY 16.3
Indicators as Weak Acids
CD-ROM Viewing Guide

Watch the Problem-Solving Skills CD-ROM demonstration "Phenolphthalein Indicator." Read Section 16.7 in your textbook and then answer the following questions.

1. Write structural formulas for the acidic and basic forms of phenolphthalein. Which form is pink? Which form is colorless? (*Hint:* Look on your CD-ROM under "Chem 4 Molecules Database.")

2. Write a reaction representing phenolphthalein

    a. as HIn, acting as a weak acid.

    b. as In⁻, acting as a weak base.

3. Phenolphthalein is useful as an indicator when the stoichiometric point of an acid base reaction occurs near what pH?

4. Pick an indicator that would be useful for the following titrations (refer to Table 16.3).

   a. hydrochloric acid and sodium hydroxide

   b. acetic acid and sodium hydroxide

   c. hydrochloric acid and ammonia

5. For fun, you can also watch the Problem-Solving Skills CD-ROM demonstration titled "Rose Petal Pigments (A Natural Indicator)."

ACTIVITY 16.4

Name _____

Buffers

Group _____

*Group Challenge Problem with
Individual Accountability*

---

1. a. You need to prepare a buffer solution with a certain pH. Decide, using Tables 15.2 and 16.4, which weak acid and base to use. The pH of your buffer should be

   (even-numbered groups) 4.5

   (odd-numbered groups) 10.0

   b. Find the ratio of conjugate acid-conjugate base concentrations that will give you the exact pH desired. Compare your results with your neighbors. Even- and odd-numbered groups should find a group consensus and write it on the chalkboard.

   (even-numbered groups) ratio

   (odd-numbered groups) ratio

2. Use the conjugate acid-base pair that you decided on in Question 1a, and assume that the concentrations of both components of your buffer are 0.10 mol/L. Calculate the pH of your buffer after the following amounts of acid and base are added to 100.0 mL of buffer solution. Assume no change in volume.

   Person A: 0.0010 mol HCl

   Person B: 0.0010 mol NaOH

   Person C: 0.00050 mol HCl

   Person D: 0.00050 mol NaOH

   I am person _____.

   Work:

STUDENT COMPANION FOR CHEMISTRY

ACTIVITY 16.5
Solubility
Guided Reading and Group Problem

Name _____

Group _____

Read the textbook Sections 16.11, 16.12, and 16.13.

1. Describe what is happening to dissolved and undissolved solute in a saturated solution, using equilibrium concepts.

2. Write the expression for K_{sp} for the following reaction: $CaF_2(s) \rightleftharpoons Ca^{2+}(aq) + 2\ F^-(aq)$.

   What substance is not included in the K_{sp} expression? Explain why it can be omitted.

3. Describe a step-by-step strategy that can be used to determine the concentration of Cd^{2+} ions in a saturated aqueous solution of cadmium carbonate. K_{sp} is 2.5×10^{-14} for $CdCO_3$.

4. Use the strategy that you developed in Question 3 to solve the problem.

5. Complete the textbook exercises assigned by your instructor.

Return to class and complete the following questions with your group members.

6. Use the letter S to represent the molar solubility of formula units in a saturated solution. For example, consider the reaction

$$AgCl(s) \rightleftharpoons Ag^+(aq) + Cl^-(aq)$$

Write the K_{sp} expression, substituting S in the place of $[Ag^+]$ and $[Cl^-]$. You can do this because each formula unit of AgCl gives rise to one Ag^+ ion and one Cl^- ion.

7. Write a K_{sp} expression as you did in Question 6, for the following compounds.
   Person A: PbI_2

   Person B: $Al(OH)_3$

   Person C: $Fe(OH)_3$

   Person D: $Cu(OH)_2$

   I was person _____.

   Expression:

   Calculate the molar solubility of your substance from its solubility constant. (Use the data table in Activity 16.6.)

ACTIVITY 16.6
Precipitates
Group Problem

Name _____

Group _____

1. A solution contains a mixture of zinc, copper, and cadmium ions in equal concentrations. A second solution containing sulfide ions is added slowly. Arrange the metal ions in the order in which they will begin to precipitate out as sulfides. *Explain your reasoning.*

2. Work Challenging Exercise 16.110 from the textbook.

DATA TABLE

Substance	K_{sp}	Substance	K_{sp}
AgCl	1.6×10^{-10}	$Cu(OH)_2$	1.6×10^{-19}
$Al(OH)_3$	1.0×10^{-33}	CuS	1.3×10^{-36}
ZnS	1.6×10^{-24}	$Fe(OH)_3$	2.0×10^{-39}
$BaCO_3$	8.1×10^{-9}	$MgCO_3$	1.0×10^{-5}
$CaCO_3$	8.7×10^{-9}	CdS	1.0×10^{-28}
PbI_2	1.4×10^{-8}		

Connection 4
What's in Our Water?

Name _____

Group _____

The chemistry behind a properly functioning swimming pool or hot tub is very important. In order to support yourself during your college years, you get a job as a swimming pool manager. You remember from chemistry class and your experiences swimming in pools that chlorine is used in water supplies because it is a very effective disinfectant. You are hired because you have a good working knowledge of the chemistry involved in pool maintenance. Your first job is to check the acidity of the water. The pH of water in a swimming pool should be between 7.2 and 7.8. During a daily sampling of the water (which consists of three samples taken from three different areas of the pool), you come up with the following data:

Chemical identity	pH	Color using bromothymol indicator solution	Temperature °C
Sample 1	6.3	yellow	29
Sample 2	8.2	blue	27
Sample 3	7.9	blue	28

Answer the following questions about the data above and how it relates to the chemistry you have learned so far. You may need reference material such as a Merck Index or the World Wide Web to determine the properties of chlorine.

1. Why do you think that people choose not to chlorinate their pools with chlorine gas instead of calcium or sodium hypochlorite?

2. Write an equilibrium equation that describes what happens when solid calcium hypochlorite reacts with water to form hypochlorous acid and slightly soluble calcium hydroxide.

STUDENT COMPANION FOR CHEMISTRY

3. Calculate the acidity constant of hypochlorous acid (HClO) in a 0.1000 m HClO solution. You determine, from the pH of a sample that the H_3O^+ is 1.34×10^{-3} mol/L. (*Hint:* Consider the steps below.)

   a. Write the proton transfer equilibrium for HClO.

   b. What are the initial concentrations of all species?

   c. Determine the change in concentration for all species.

   d. What is the concentration of each species at equilibrium?

   e. Now, using this data, determine the K_a of hypochlorous acid.

4. Why is it dangerous to store pool chlorine (in the form of HOCl) near materials that can burn easily? Consider redox reactions in your answer.

5. Discuss the overall "quality" of the water from the pool data. Would this be safe water for the public to swim in?

Chapter 17

The Direction of Chemical Change

Activity 17.1   The Laws of Thermodynamics—Review Worksheet, p. 279
Activity 17.2   Determining Spontaneity—Worksheet, p. 281
Activity 17.3   Group Demonstration—Viewing Guide, p. 283
Activity 17.4   Group Problem—Worksheet, p. 285
Activity 17.5   Group Challenge Problem—Worksheet, p. 287
Activity 17.6   Equations—Review Worksheet, p. 289

ACTIVITY 17.1
The Laws of Thermodynamics
Review Worksheet

Name _____

Group _____

1. For each statement or equation, write in the correct corresponding numbered law of thermodynamics.

    a. _____ The entropy of a perfect crystal is 0.

    b. _____ The total amount of energy in the universe is constant.

    c. _____ As time passes, entropy in a system increases.

    d. _____ $\Delta S_{surr} = \Delta H/T$

2. Describe the changes in the surroundings that result from an exothermic process.

3. Which member of the following pairs has the highest molar entropy at 298 K?

    a. $CO_2(g)$ or $CO_2(s)$

    because:

    b. one mole of $Cl_2(g)$ at 1.0 atm or one mole of $Cl_2(g)$ at 1×10^{-3} atm

    because:

STUDENT COMPANION FOR CHEMISTRY

c. Mg(s) or Mg(l)

   because:

4. What information would be required to determine the molar $\Delta S°$ of a phase change?

5. Use Appendix 2A to calculate the ΔS at 25°C for the following reaction:

   Solid aluminum oxide reacts with hydrogen gas to produce aluminum metal and gaseous water.

6. If NO(g) and O_2(g) form NO_2(g), what sign would you expect $\Delta S°$ to have? Explain.

7. Calculate $\Delta S°$ for this reaction. (Was your prediction accurate?)

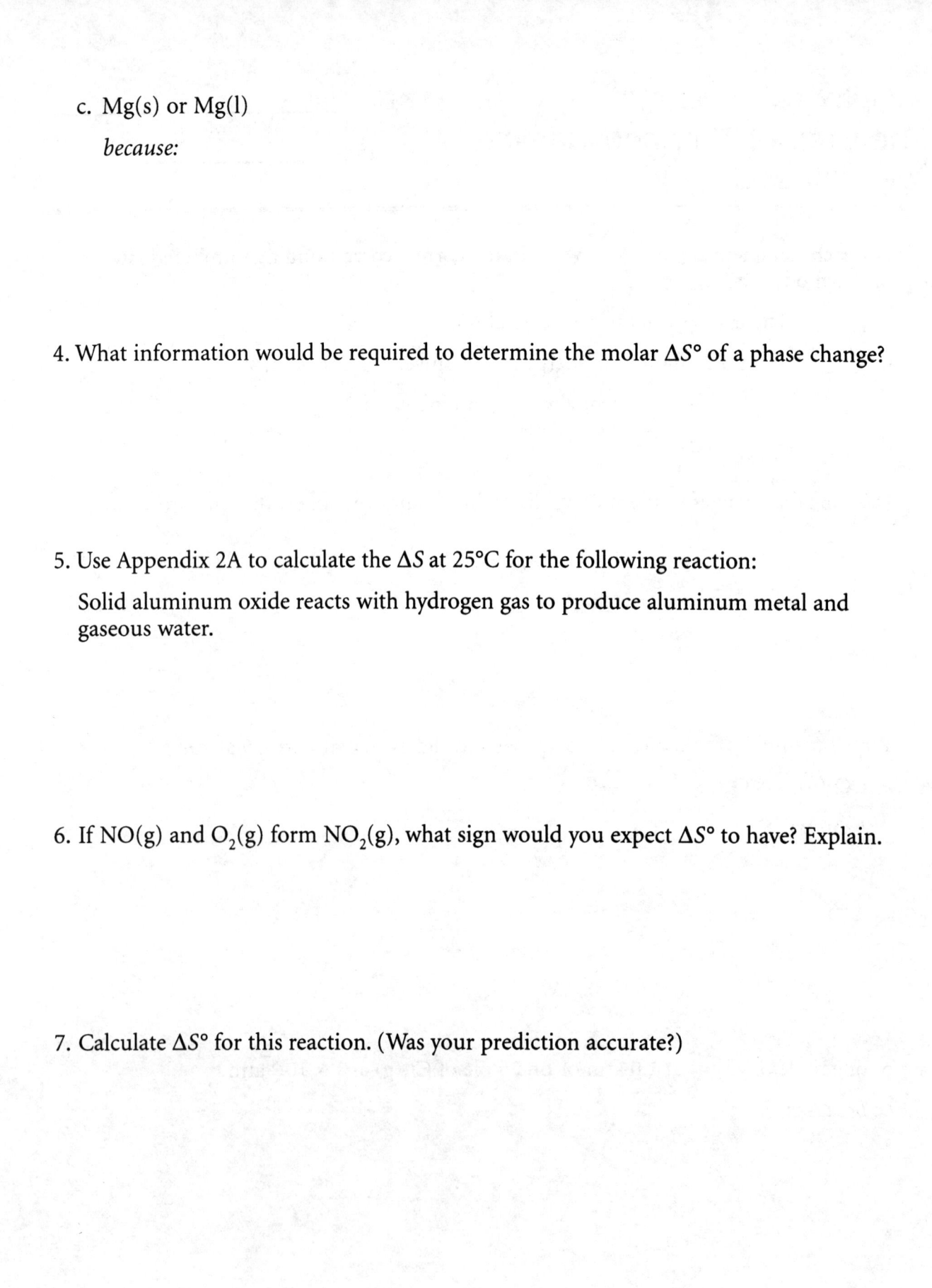

ACTIVITY 17.2
Determining Spontaneity
Worksheet

Name _____

Group _____

To understand whether a reaction will be spontaneous or not, you need to understand a few concepts. Respond to the following questions.

1. Explain, using your own terminology, what is meant by **entropy**.

2. If a system is in a state of equilibrium, the ΔG should be _____.

3. If the system does work, the free energy _____ and the process is considered to be spontaneous.

4. Complete the following chart.

If w = _____	Work done?	ΔG = _____	Describe the reaction
0		0	
+		−	
−		+	

5. Under what conditions are the values of ΔG and ΔH for a process very close in sign and magnitude?

STUDENT COMPANION FOR CHEMISTRY

6. Use the formula $\Delta G = \Delta G° + 2.303RT \log Q$ and Appendix 2A to determine the ΔG for the following reaction: $2\ HI(g) \rightleftharpoons H_2(g) + I_2(s)$

7. Use the chart from Question 4 to discuss the spontaneity of the reaction in Question 6.

8. Now consider ΔH and its importance in determining spontaneity. Complete this chart.

ΔS	ΔH	Energy change?	Entropy change of the universe?	Spontaneous reaction?
+	−	exothermic	+	Yes
+	+			
−	−			
−	+			

9. Describe a strategy that will help you determine the spontaneity of a reaction. What information will you need to consider to determine this?

ACTIVITY 17.3
Group Demonstration
Viewing Guide

Name _____

Group _____

Watch the demonstration done by your instructor, then answer the following questions.

1. What are the reactant(s) used in this demonstration?

2. What are the product(s) formed in this demonstration?

3. Write a balanced chemical equation for the reaction performed in the demonstration.

4. Discuss and explain the entropy changes that occurred in the flask.

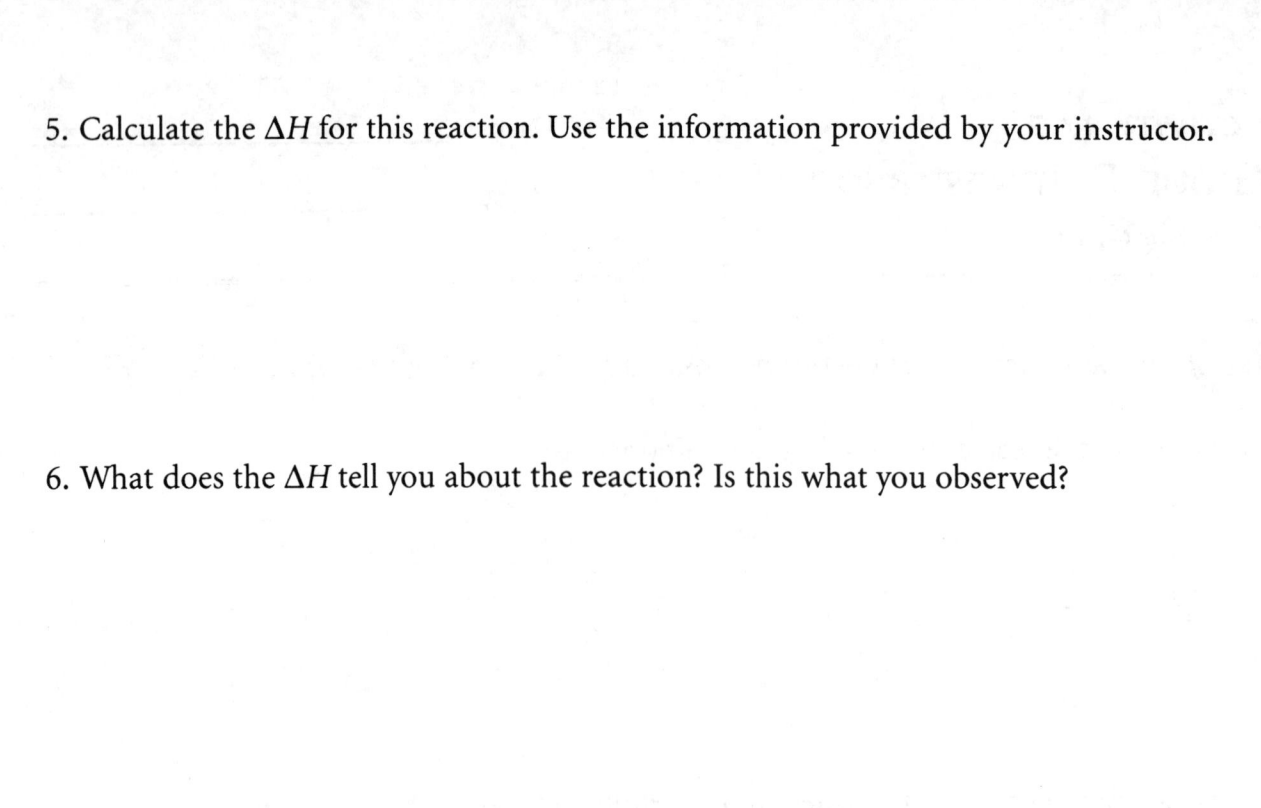

5. Calculate the ΔH for this reaction. Use the information provided by your instructor.

6. What does the ΔH tell you about the reaction? Is this what you observed?

7. How could you determine whether this reaction is spontaneous? Based on your discussion, do you think the reaction is spontaneous?

ACTIVITY 17.4
Group Problem
Worksheet

Name _____

Group _____

Yeast is a one-celled organism that has the ability to ferment the sugar in various fruits to form alcohol. One example is the fermentation of grapes to produce wine. The first of many steps involves an aqueous glucose solution that reacts with oxygen to form ethanol, carbon dioxide, and water.

1. Balance the chemical reaction for this process.

$$C_6H_{12}O_6(s) + O_2(g) \rightarrow C_2H_5OH(l) + CO_2(g) + H_2O(l)$$

2. Calculate the ΔS for the preceding reaction.

Substance	ΔH_f° (kJ/mol)	S_m° (J/K mol)	ΔG_f° (kJ/mol)
Glucose (s)	−1268	212	−910
Ethanol (l)	−278	161	−175
Carbon dioxide (g)	−394	214	−394
Water (l)	−285	70	−237
Oxygen (g)	0	205	0

3. Would this reaction be considered spontaneous? Explain.

4. Assuming that you begin with 1.40 kg of glucose, how many grams of ethanol and of carbon dioxide could be produced?

ACTIVITY 17.5
Group Challenge Problem
Worksheet

Name _____

Group _____

Hydrogen peroxide is used commercially for a variety of purposes. A 3% solution is typically used as a topical antiseptic. The H_2O_2 decomposes into water and oxygen gas. It is the oxygen gas that is responsible for the "bubbling" action that is produced when the peroxide comes in contact with a wound. Hydrogen peroxide as strong as 90% can be used in rocket propulsion. There are a variety of ways to produce hydrogen peroxide. Two possible ways are:

Method 1: $2\ H_2O(l) + O_2(g) \rightarrow H_2O_2(l)$

Is water being oxidized or reduced in this method? _____

Method 2: $H_2(g) + O_2(g) \rightarrow H_2O_2(l)$

Is oxygen being oxidized or reduced in this method? _____

1. As a group, develop a problem-solving strategy that would allow you to determine which method would be the most thermodynamically economical at producing hydrogen peroxide commercially. (*Hint:* Consider what the term *economical* means.)

2. Use your strategy to determine which method is most economical and *explain your reasoning*. Some information that might prove helpful are the $\Delta G°$ values for hydrogen peroxide (l) and water (l): −120 kJ/mol and −237 kJ/mol respectively. You already know the $\Delta G°$ values for hydrogen and oxygen gases.

3. Explain how a 3% solution would be made from a 90% solution.

4. As homework, use the World Wide Web to find out how H_2O_2 is produced industrially. Compare what you learned in this problem to what you found on the Web.

ACTIVITY 17.6

Equations

Review Worksheet

Name _____

Group _____

1. *Symbols and meanings* Explain what is meant by each of the following symbols.

   a. ΔH_r° _____

   b. ΔH_f° _____

   c. ΔG_r° _____

   d. ΔG_f° _____

   e. ΔS_r° _____

   f. S_m° _____

   g. Σ _____

   h. ΔU _____

   i. w _____

   j. K _____

STUDENT COMPANION FOR CHEMISTRY

2. Create a problem that requires the use of each of the following equations.

   a. $\Delta G_r° = \Delta H_r° - T\Delta S_r°$

   b. $\Delta G = RT \ln K$

3. Solve either your own problems that you created above, or exchange papers with someone else. Either way, it will help you to review and use these equations.

Chapter 18

Electrons in Transition: Electrochemistry

Activity 18.1　Balancing Redox Reactions—Worksheet, p. 293
Activity 18.2　Galvanic Cells—Worksheet and CD-ROM Demonstration, p. 295
Activity 18.3　Weird Cells: Cell Potentials, Free Energy, and Equilibrium Constants—Group Problem, p. 299
Activity 18.4　The Nernst Equation—Guided Reading, p. 301
Activity 18.5　Group Challenge Problem, p. 305

ACTIVITY 18.1
Balancing Redox Reactions
Worksheet

Name _____

Group _____

Watch the Problem-Solving Skills CD-ROM demonstration "Reaction of Aluminum and Iron Oxide: The Thermite Reaction." Answer the following questions.

1. Write a skeletal (unbalanced) chemical equation for this reaction. Which species is oxidized and which is reduced?

2. Write half-reactions for the thermite reaction. Which half-reaction represents oxidation? Which represents reduction? Balance the chemical equation, using the half-reactions.

3. Write your own strategy for balancing oxidation-reduction reactions in acidic and in basic solutions.

4. Use your strategy to balance the following skeletal equations.

   a. in acidic solution: $MnO_4^-(aq) + H_2SO_3(aq) \rightarrow Mn^{2+}(aq) + HSO_4^-(aq)$

   b. in basic solution: $MnO_4^-(aq) + S^{2-}(aq) \rightarrow S(s) + MnO_2(s)$

ACTIVITY 18.2
Galvanic Cells
Worksheet and CD-ROM Demonstration

Name _____

Group _____

Read textbook Sections 18.3 through 18.5, and watch the Problem-Solving Skills CD-ROM demonstration "A Galvanic Cell."

1. Draw a picture of the cell used in this demonstration.

2. What happens when

   a. the electrodes touch?

   b. the solutions are mixed?

STUDENT COMPANION FOR CHEMISTRY

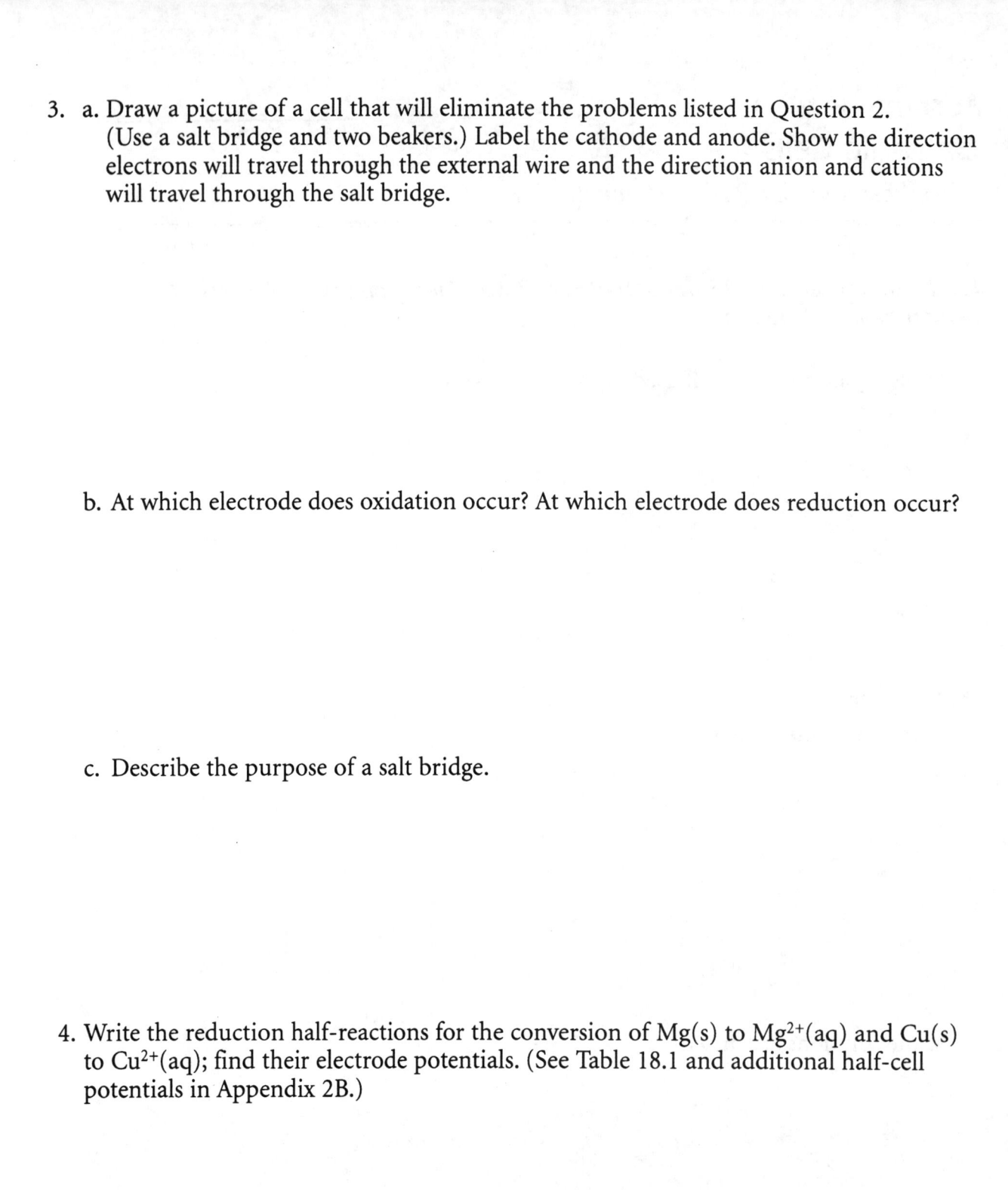

3. a. Draw a picture of a cell that will eliminate the problems listed in Question 2. (Use a salt bridge and two beakers.) Label the cathode and anode. Show the direction electrons will travel through the external wire and the direction anion and cations will travel through the salt bridge.

b. At which electrode does oxidation occur? At which electrode does reduction occur?

c. Describe the purpose of a salt bridge.

4. Write the reduction half-reactions for the conversion of Mg(s) to Mg^{2+}(aq) and Cu(s) to Cu^{2+}(aq); find their electrode potentials. (See Table 18.1 and additional half-cell potentials in Appendix 2B.)

5. Combine these two reactions into an overall spontaneous reaction and calculate the cell potential.

6. Write the cell notation for the reaction.

ACTIVITY 18.3
Weird Cells: Cell Potentials, Free Energy, and Equilibrium Constants
Group Problem

Name _____

Group _____

---

1. Pick two half-reactions. (You can be creative, but they must be in the electrochemical series presented in Table 18.1 or Appendix 2B in your textbook.) Pass these reactions on to a neighboring group.

2. Using the half-cell provided by the other group, design a galvanic cell.

3. Write the half-reactions and the overall reaction for the cell.

4. Calculate the standard potential ($E°$) for the cell.

5. Calculate the free energy (ΔG), and the equilibrium constant for the overall cell reaction.

6. Pass your work back to the originator of this weird cell.

7. Check over the calculations done by the other group on the weird cell your group designed and discuss any corrections with the group you exchanged weird cells with.

   Work:

ACTIVITY 18.4
The Nernst Equation
Guided Reading

Name _____

Group _____

In the last part of the nineteenth century, the German chemist H.W. Nernst found the relationship between the voltage of a cell and the conditions under which it operates. The voltage of a cell, under standard conditions, is called $E°$. Standard conditions are considered to be 25°C, 1.00 M, and approximately 1 atm of pressure. If conditions are other than standard, the voltage is called E. Using this, Nernst came up with the equation that bears his name:

$$E = E° - RT/nF \ln Q$$

1. Complete the following chart related to the symbols in this equation.

Symbol	Refers to	(Value) and units
	universal gas constant	8.31 J/mol K
K	standard temp. condition	
n		
	conversion factor	96,465 J/mol
ln		2.30 log (using base 10 log)

Combine the values for the symbols $RT/F \times 2.30$ (from above) and rewrite the equation, using standard temperature.

$$E =$$

Read Section 18.10 in your textbook and answer the following questions.

2. Develop a strategy to solve problems by using the Nernst equation, considering the following:

   a. Under which conditions (standard or nonstandard) will you use the Nernst equation to calculate the cell potential?

   b. What information do you need to know to solve for nonstandard cell potentials? for concentrations of cell components?

STUDENT COMPANION FOR CHEMISTRY

3. Complete the homework problems assigned by your instructor

4. As an entire group, use the Nernst equation to solve the following problem.

   What would be the voltage of a cell using Ni | Ni^{2+} (2.50 mol/L) for one half-cell and Ag | Ag$^+$ (0.100 mol/L) for the other half-cell? Assume pressure and temperature to be standard values.

5. Individually, determine the voltage of one of the following cells from the given information. You will compare your individual results with those of your group members. Assume all cells are at 25°C and standard pressure.

   Person A: Pb | Pb^{2+} (0.485 mol/L) || Sn^{4+} (0.652 mol/L) | Sn^{2+} (0.346 mol/L)

   Person B: Ni | Ni^{2+} (1.00 mol/L) || Cu^{2+} (1.00 mol/L) | Cu

   Person C: Fe | Fe^{2+} (0.720 mol/L) || Ag$^+$ (0.785 mol/L) | Ag

   I was person _____ for the individual portion.

   Work:

6. Sketch a schematic diagram for the cell from Question 5 that produced the *highest* voltage.

7. Nernst was awarded the Nobel Prize in chemistry for his work! Suggest some reasons why his work was considered so important.

ACTIVITY 18.5
Group Challenge Problem

Name _____

Group _____

Work this problem with your group members. Assume all temperatures are 25°C.

A common reference electrode consists of a silver wire coated with AgCl(s) and immersed in a 1.00 M KCl solution. The half-reaction for this electrode is

$$AgCl(s) + e^- \rightarrow Ag(s) + Cl^-(aq) \qquad E° = +0.22 \text{ V}$$

Using the standard potentials listed in your textbook and the silver-silver chloride reference electrode, set up a spontaneous cell to measure the potential for the following net reaction:

$$Ag^+(aq) + Cl^-(aq) \rightarrow AgCl(s)$$

1. Write the half-reactions that occur in this cell at the

   anode.

   cathode.

2. Calculate the $E°_{cell}$, and $\Delta G°$ for the net reaction.

STUDENT COMPANION FOR CHEMISTRY

3. Calculate K_{sp} for AgCl. (Assume standard conditions and 25°C.)

4. Using the Le Chatelier principle, explain how the following conditions would change the cell potential for the net reaction in Question 1.

   increasing the concentration of silver ions

   decreasing the concentration of chloride ions

Connection 5
Electric Vehicles

Name _____

Group _____

Automobiles that are powered by fuel cells and electric vehicles are very efficient and produce no pollution. And ideal battery for an electric car should:

- have a specific energy
- have a high energy density
- have a long range of operation per charge
- have a long life cycle
- be environmentally safe when disposed of
- be inexpensive to operate

1. Explain what specific energy and density mean.

Some possible batteries for use in electric vehicles are the lead-acid cell, nickel-cadmium cell (nicad), nickel-metal hydride, lithium-ion cell and lithium-polymer cell, zinc-air cell, and sodium-sulfur cell. These cells are described in Connection 5.

2. Which of these batteries would be the most "ideal," based on the criteria listed above? Explain.

3. All of these batteries are secondary cells. Explain the differences between a primary cell, secondary cell, and fuel cell. Why is a secondary cell used in a car instead of a primary cell?

STUDENT COMPANION FOR CHEMISTRY

4. The nicad cell is used to power electronic equipment. Write the net and half-reactions for this cell.

   Net reaction:

   Anode half-reaction:

   Cathode half-reaction:

5. Calculate the standard cell potential, equilibrium constant and standard free energy for the nical cell. Compare your standard cell potential to the cell potential given in Table 18.2 of your textbook. Why are these values different?

6. Write the half-reactions that occur at the anode and cathode when this battery is recharged.

   Anode:

   Cathode:

Chapter 19

The Elements: The First Four Main Groups

Activity 19.1   Properties of H, Ca, and Mg—CD-ROM Demonstration Viewing Guide, p. 311
Activity 19.2   Carbon Allotropes—Worksheet, p. 313
Activity 19.3   Main-Group Elements—Review Chart, p. 315
Activity 19.4   Elements in Groups 1, 2, 13, and 14 and Their Uses—Worksheet, p. 317

ACTIVITY 19.1
Properties of H, Ca, and Mg
CD-ROM Demonstration Viewing Guide

Name _____

Group _____

View the demonstrations on the Problem-Solving Skills CD-ROM titled "Properties of Hydrogen" and "Reaction of Magnesium and Oxygen." Answer the following questions, either as you watch or as a review of the information presented. Your text may also provide useful information.

1. List three properties of hydrogen gas.

2. Sketch a molecule of hydrogen gas.

3. Explain in terms of molecular structure and bonding why hydrogen gas has such a low melting point (−260°C) and boiling point (−253°C).

STUDENT COMPANION FOR CHEMISTRY

4. Compare, using examples, **ionic hydrides, molecular hydrides,** and **metallic hydrides.**

5. What is the source of most of the calcium on Earth?

6. Calcium ions are commonly found in water supplies. Why are these ions so common and what effects are related to the presence of calcium?

7. Calcium is essential to humans. Bone is composed of protein fibers, water, and other minerals. One of the most important is hydoxyapatite. What is the formula of this compound?

8. Which type of fire extinguisher can you use to extinguish a magnesium fire: water or CO_2? Why?

ACTIVITY 19.2
Carbon Allotropes
Worksheet

Name _____

Group _____

1. Label the following diagrams as graphite or diamond.

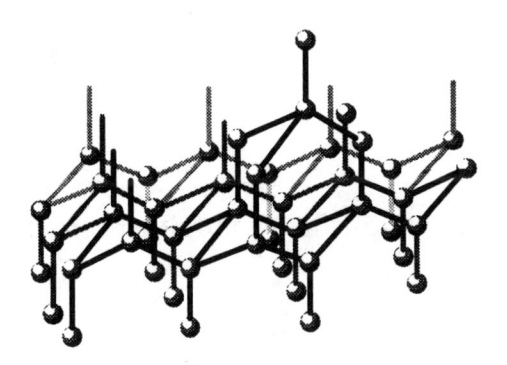

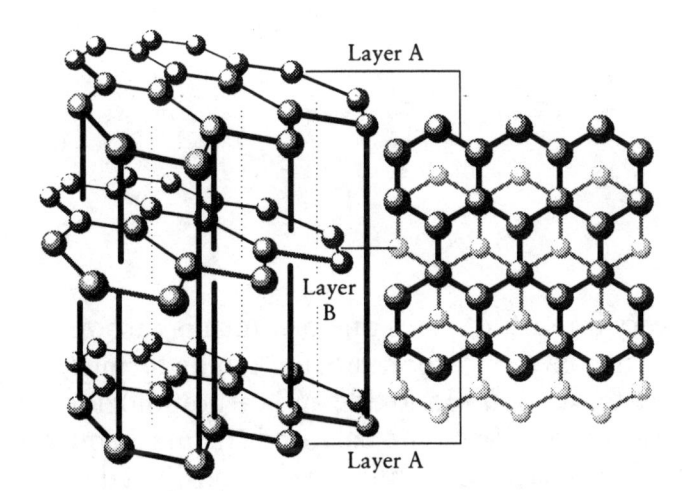

2. Explain how these structures fit the definition of **allotropes.**

3. Which allotrope would best be used as a lubricant? Explain.

4. Discuss how diamond could be made from graphite.

5. Explain why some properties of carbon are similar to those of silicon and others are different. Use examples in your explanation.

6. Carbon is also a component of several very important gases one of which is carbon dioxide. When carbon monoxide reacts with oxygen, carbon dioxide can be formed. The $\Delta H_f°$ for oxygen is 0. The $\Delta H_f°$ values for carbon monoxide and carbon dioxide are −110.5 kJ/mol and −393.5 kJ/mol, respectively. Write the balanced chemical equation for this reaction and calculate the $\Delta H°$ of this reaction. Would this be considered a spontaneous reaction? Why or why not?

ACTIVITY 19.3
Main-Group Elements
Review Chart

	Group 1	*Group 2*	*Group 13*	*Group 14*
Symbols of members				
Valence electron configuration				
Name of group				
Describe the reaction with water typical of this group				
Trend in electronegativity				
Describe elements that are unusual or do not follow group trends				
Trend in atomic radius				
Industrial or commercial uses				
Miscellaneous information				

STUDENT COMPANION FOR CHEMISTRY

ACTIVITY 19.4

Elements in Groups 1, 2, 13, and 14 and Their Uses
Worksheet

Name _____

Group _____

With your group, periodic table, and/or perhaps your text, match the correct element symbol from Groups 1, 2, 13, or 14 with the appropriate use *or* description.

_____ 1. Diatomic nonmetal; most abundant element in the universe

_____ 2. Hardest Group 1 element

_____ 3. Group 13 nonmetal that forms acidic oxides

_____ 4. Has half-filled valence shell; used in semiconductors

_____ 5. Forms clusters of 12 atoms and has several allotropes

_____ 6. The only alkali metal that reacts with nitrogen to form nitride

_____ 7. Forms acidic oxides and is basis of organic chemistry

_____ 8. Main sources are sylvite and carnallite minerals; used in fertilizer

_____ 9. Has three allotropes: graphite, diamond, and fullerene

_____ 10. Used to make washing soda, baking soda, and soda ash

_____ 11. Most abundant metallic element in Earth's crust; obtained by Hall process

_____ 12. Used as a very lightweight alloy component; burns in N_2, O_2 and CO_2

_____ 13. Derived from galena ore; density makes it a good radiation shield

_____ 14. Monoxide of this element has the highest bond enthalpy of all molecules

_____ 15. Used with liquid oxygen to fuel space shuttle main rocket engines

_____ 16. The only strongly electropositive metal that will not form a saline hydride

_____ 17. The most reactive nonradioactive metal

_____ 18. Component of quicklime; important to metallurgy

_____ 19. Group 2 element that will not react with water

_____ 20. Metalloid found in quartz, quartzite, and cristobalite

STUDENT COMPANION FOR CHEMISTRY

Chapter 20

The Elements: The Last Four Main Groups

Activity 20.1 Properties of Phosphorus, Oxygen, and Nitrogen—CD-ROM Demonstration Viewing Guide, p. 321

Activity 20.2 Group Trends—Group Problem, p. 325

Activity 20.3 Groups 15–18—Review Chart, p. 327

Activity 20.4 Modeling and Diagramming Group 15–18 Elements, p. 329

ACTIVITY 20.1
Properties of Phosphorus, Oxygen, and Nitrogen
CD-ROM Demonstration Viewing Guide

Name _____

Group _____

View the demonstration on the Problem-Solving Skills CD-ROM titled "Reaction of Phosphorus and Oxygen." Answer the following questions as you watch the demonstration. Your text also provides useful information.

1. At what temperature does nitrogen gas become a liquid?

2. Sketch the Lewis structure of an N_2 molecule. Explain at least two properties of N_2 in terms of its bonding.

3. How is liquid nitrogen used commercially?

4. Describe the two allotropic forms of phosphorus. How are they similar and how are they different?

STUDENT COMPANION FOR CHEMISTRY

5. Draw the structure of a P₄ molecule.

6. a. Write the reaction that occurs when phosphorus atoms react with excess oxygen. What is formed when this phosphorus product is then combined with water?

   b. Write the reaction that occurs when phosphorus atoms are combined with a limiting reactant, oxygen. What is formed when this phosphorus product is then combined with water?

7. Discuss the role of phosphorus in fireworks and explosives.

8. Suppose you were given the following materials: red Phosphorous, white phosphorus, powdered glass, cardboard, and sulfur. Design a match that could be used to safely light a Bunsen burner.

ACTIVITY 20.2
Group Trends
Group Problem

Name _____

Group _____

Your instructor will give you a set of nine cards of fictitious elements. The cards contain several pieces of information that you may use to organize them in a periodic table. When you have decided where each "element" belongs, attach it below or onto a separate sheet of paper.

1. What characteristic(s) was/were most helpful in determining where "elements" belonged?

2. Which "elements" were gases at room temperature and pressure?

3. For each of the elements in Question 2, identify which *real* element would *most* resemble the "fictitious" descriptions.

STUDENT COMPANION FOR CHEMISTRY

4. What other physical and chemical properties would be helpful in the identification of these elements? (*Hint:* Consider trends that typically occur either across a period or down a group.)

ACTIVITY 20.3
Groups 15–18
Review Chart

Name _____

Group _____

	Group 15	Group 16	Group 17	Group 18
Symbols of members				
Valence electron configuration				
Name of group				
Describe the reaction with water typical of this group				
Common compounds formed with hydrogen				
Trend in electronegativity				
Trend in atomic radius				
Trend in ionic radius				
Miscellaneous information				

STUDENT COMPANION FOR CHEMISTRY

ACTIVITY 20.4
Modeling and Diagramming Group 15–18 Elements

Name _____

Group _____

Use the atomic modeling kits from your instructor, if provided. For each of the following,

a. name the compound.

b. identify the group that the central atom represents.

c. draw the Lewis dot structure and use formal charges to find the best structures.

d. label all bond angles.

1. HNO_3

2. H_2SO_3

3. $HClO_3$

4. XeO_3

STUDENT COMPANION FOR CHEMISTRY

5. Discuss any trends that occur as you move from a Group 15 oxoacid to a Group 17 oxoacid. Are the characteristics of XeO$_3$ consistent with this trend? Explain.

Chapter 21

The *d* Block: Metals in Transition

Activity 21.1	The Physical and Chemical Properties of Transition Metals—Guided Reading and CD-ROM Demonstration, p. 333
Activity 21.2	Isomers—Group Model Building, p. 337
Activity 21.3	Crystal Field Theory: Ligands and Color—Group Problem and Lecture Demonstration, p. 339
Activity 21.4	Group Challenge Problem, p. 341

ACTIVITY 21.1

The Physical and Chemical Properties of Transition Metals

Name _____

Group _____

Guided Reading and CD-ROM Demonstration

Trends in Physical Properties

Read Section 21.1.

1. Describe the periodic trend in atomic radii down each transition metal group.

2. a. Using Figure 21.4 and the Problem-Solving Skills CD-ROM curve-fitter, plot the atomic radius for each transition metal in the fourth period on the *y*-axis and the atomic number of each transition metal on the *x*-axis.

   b. Describe and explain your observed periodic trend.

3. Why do some transition metals such as iron, cobalt, and nickel make good permanent magnets?

STUDENT COMPANION FOR CHEMISTRY

4. Use Table 21.1 and your CD-ROM curve-fitter to plot the melting point trend for transition metals in the fourth period. Describe the observed trend below.

Chemical Properties

Read Section 21.2 and watch the Problem-Solving Skills CD-ROM demonstration "Oxidation States of Ammonium Vanadate."

5. Describe the periodic trends in multiplicity of oxidation states.

6. a. Record the colors and oxidation states for vanadium that you observed in the CD-ROM demonstration.

   b. Which of these oxidation states are the most common ones for vanadium?

7. Explain why MnO_4^- and $Cr_2O_7^{2-}$ are good oxidizing agents, whereas $FeCl_2$ is a good reducing agent.

Read Sections 21.3 and 21.4.

8. Record any facts about transition metals that you find interesting. Return to class and design a 10-question "transition metal trivia quiz" with the help of your fellow group members. Exchange quizzes with a neighboring group. Complete your neighbor's quiz, pass it back, and correct. Discuss any corrections together.

ACTIVITY 21.2

Isomers

Group Model Building

Name _____

Group _____

1. Read Section 21.6 as homework and fill in the following table.

Structural isomers	Example/description
ionization	
hydrate	
linkage	
coordination	
Stereoisomers	*Example/description*
geometrical	
optical	

2. Build models for all the stereoisomers of the following complexes. Sketch and label optical and geometrical isomers.

   a. $[PtCl_2(NH_3)_2]$ (a square planar complex)

STUDENT COMPANION FOR CHEMISTRY

b. $[Co(en)_3]^{3+}$

c. $[Cr(ox)_2(H_2O)_2]^-$

ACTIVITY 21.3

Crystal Field Theory: Ligands and Color

Group Problem and Lecture Demonstration

Name _____

Group _____

1. Order the following ligands, from weak-field to strong-field: H_2O, NH_3, en, CN^-.

2. Using Figure 21.40, predict the colors of the following nickel complexes.

   a. $[Ni(H_2O)_6]^{2+}$

   b. $[Ni(NH_3)_6]^{2+}$

   c. $[Ni(CN)_4]^{2-}$

3. Watch the lecture demonstration and record your observations below (dmg is dimethylglyoxime, $C_4H_8N_2O_2$, a bidentate ligand).

Complex	Color
$[Ni(H_2O)_6]^{2+}$	
$[Ni(NH_3)_6]^{2+}$	
$[Ni(H_2O)_4(en)]^{2+}$	
$[Ni(H_2O)_2(en)_2]^{2+}$	
$[Ni(en)_3]^{2+}$	
$[Ni(dmg)_2]$	
$[Ni(CN)_4]^{2-}$	

4. How well do your predictions match what you observed? Can you explain your observations?

5. List the ligands used in this demonstration, from low-field to high-field. Base this list on your observations.

ACTIVITY 21.4
Group Challenge Problem

Name _____

Group _____

Calamine lotion, used to relieve skin irritations, contains zinc and iron oxides. The amount of zinc oxide present can be determined as follows:

- A 1.00-g sample of calamine lotion is dried, dissolved in acid, and diluted to 250.0 mL with distilled water.

- KF is added to 10.0 mL of the dissolved and diluted calamine. The KF reacts with the Fe^{3+} ions, forming a complex that will not react with EDTA.

- The pH of the resulting solution is adjusted and it is titrated with 38.0 mL of 0.0120 M EDTA(aq) to the end point.

Calculate the mass percentage of zinc oxide in the original sample of calamine lotion. *Hint:* The mole ratio of Zn^{2+} ions to EDTA in the titration is 1:1.

Chapter 22

Nuclear Chemistry

Activity 22.1   Nuclear Particles—Guided Reading, p. 345
Activity 22.2   Determining Radiation Exposure and Effects—Worksheet, p. 349
Activity 22.3   Calculating Half-Lives—Worksheet, p. 351
Activity 22.4   Fission versus Fusion—Group Challenge Problem, p. 353
Activity 22.5   Nuclear Chemistry Concept Map, p. 355

ACTIVITY 22.1

Nuclear Particles

Guided Reading

Name _____

Group _____

Read sections 22.1, 22.2, and 22.3. Answer the following questions individually and bring them with you to complete the group component.

1. Distinguish between the terms **nucleon** and **nuclide**. Use examples in your distinction.

2. Describe your understanding of the term **radiation**.

3. Complete the following chart.

Decay component	Mass	Charge	Notation used	Atomic location
alpha (α)				
beta (β)				
gamma (γ)				
positron				
proton				
electron				
neutron				

STUDENT COMPANION FOR CHEMISTRY

4. a. Which elements are most likely to decay or give off α radiation?

b. Which elements are most likely to decay or give off β radiation?

c. When does the emission of γ radiation occur?

5. Using your text as a guide, write a balanced nuclear equation for each of the following events.

a. α emission by $^{239}_{93}Np$

b. β emission by $^{30}_{13}Al$

c. positron emission by $^{99}_{43}Tc$

d. electron capture by $^{55}_{26}Fe$

e. β+ emission by $^{245}_{95}Am$

6. Complete the problems assigned by your instructor.

Complete Question 7 with your group after you have each finished answering Questions 1–6. Assign each group member a letter: A, B, C, or D. Write in the name of the person assigned to each letter.

A _____   B _____

C _____   D _____

7. The naturally occurring radioactive nuclide U-232 decays through a series of steps. Write the equation for each step as assigned. Determine the nuclide that is the final product of the steps you write. Notice that, except for A, each person will need the result of the previous person's work.

Begin with U-232

Person A: α β⁻ β⁻

Person B: α α α

Person C: α α β⁻ β⁻

Person D: α β⁻ β⁻

The final product is _____. Is this a stable product? If yes, explain why. If not, explain why not and what would need to happen for it to become stable.

ACTIVITY 22.2
Determining Radiation Exposure and Effects
Worksheet

Name _____

Group _____

Adam Averageperson was having difficulties with hair loss and decided to visit his local physician to determine the cause. As he was sitting in the waiting room, he noticed an article in a popular news magazine regarding radiation and the possible effects of exposure. The article contained a survey that could estimate radiation exposure. It stated that most people receive 20% of their total radiation from their own bodies, 30% from cosmic rays, 40% from radon exposure, and 10% from medical diagnoses.

Adam determined that he had approximately 145 mrem of background radiation, based on where he lived and the composition of his home. He had received 3 chest x-rays (10 mrem) when he had pneumonia last fall. In addition, he visited his dentist twice and was exposed to two x-rays at each visit (5 mrem per x-ray). He lives approximately 50 miles from a nuclear power plant (1 mrem). Adam traveled once by plane to visit his grandparents (2 mrem).

The article went on to state that the typical person is allowed 170 mrem/yr and up to 5 rem for occupational exposure. Observable damage occurs at approximately 20 rem/yr according to the study cited in the magazine.

1. What is meant by the term **rad**?

2. How is a rad different from a **rem**?

3. a. Beta and gamma radiation exert similar damage per rad but alpha radiation exerts _____ times that damage. Why?

STUDENT COMPANION FOR CHEMISTRY

   b. To compensate for the differences in dosage per rad, a dose equivalent is the modified method of measurement. This is measured in units called _____, where 1 Sv = 100 rem.

4. Calculate Adam's exposure in mrem, rem, and Sv.

5. Discuss whether Adam's hair loss is likely to be caused by his radiation exposure. If so, why? If not, why not?

6. Discuss the penetrating power of alpha, beta, and gamma decay. In each case, what type of shield could be used to prevent penetration into tissue?

ACTIVITY 22.3
Calculating Half-Lives
Worksheet

The half-life of a radioactive element is the time it takes for one-half of the nuclei in a sample to decay. The rate of decay can be determined by multiplying the number of radioactive nuclei (N) by k (a decay constant).

1. What relationship exists between N and the rate of decay?

2. The half-life can be determined by using the following equation:

$$t_{1/2} = \ln 2 / k$$

Sketch a rough graph that depicts the decay of any radioactive substance. Label the *x*-axis "Time" and the *y*-axis "Fraction of material remaining."

3. Isotopic dating involves the use of a half-life of a radioactive isotope. Explain how the date of an object can be estimated with carbon-14 dating.

4. A bone is discovered in an ancient cave. Carbon-14 dating shows that it contains one-fourth of the ratio of carbon-12 to carbon-14 than a modern bone contains. How old is this bone estimated to be? The half-life of carbon-14 is 5730 years.

5. Explain why carbon-14 dating loses its accuracy if an object is older than 20,000 years.

6. Tritium can also be used to date recent materials, especially wines because it has a half-life of 12.2 years. A bottle of wine is shown to have 12.5% of the tritium activity that a new bottle of wine would have. How old is this wine estimated to be? Is this an accurate estimation? Explain.

7. Hospitals often use radioisotopes as a source of ionizing radiation in the treatment of cancer. If a hospital starts with a 2000-mg supply of cobalt-60, which has a half-life of 5 years, how much of this isotope would need to be purchased to replenish the supply lost after 15 years?

If a patient were treated with 10 mg of cobalt-60, how long would it take for this amount to decay to less than 1 mg?

8. Using information from your text, write a half-life question that could be used on a quiz or an exam for this class. (Provide a worked-out solution, so that the answer can be checked.)

ACTIVITY 22.4
Fission versus Fusion
Group Challenge Problem

Name _____

Group _____

Information to consider:

1 neutron = 1.0087 u   Strontium-90 = 89.908 u
1 proton = 1.0078 u   helium-4 = 4,0026 u
1 deuteron = 2.0135 u   uranium-235 = 235.04 u
1 tritium = 3.01550 u   xenon-143 = 143.003
1 u = 1.6605 × 10⁻²⁷ g

$\Delta m = \Sigma m_{prod} - \Sigma m_{react}$

1. Develop a group problem-solving strategy that you could use to determine the amount of energy (in joules) that could be released when a fission or fusion reaction occurs.

2. Calculate the energy released (in joules) when a 10.0-g uranium-235 sample undergoes fission to produce strontium-90, xenon-143, and some neutrons.

3. Calculate the energy released (in joules) when the same mass of hydrogen-2 undergoes fusion in such a way that six atoms of deuterium produce helium-4, two protons, and two neutrons.

STUDENT COMPANION FOR CHEMISTRY

4. Which process produces the most energy from the same initial mass?

5. Discuss the advantages and disadvantages of fission versus fusion.

6. Fusion involves the joining of small molecules to form medium-sized molecules. Fission involves the splitting of large molecules to form medium-sized molecules. How can both of these processes give off energy?

Activity 22.5
Nuclear Chemistry Concept Map

Name _____

Group _____

Develop a concept map, beginning in the center of this page (or use a larger page, if necessary). Include definitions, explanations, equations, symbols, etc.—anything that you have learned about nuclear chemistry.